AF777001

METHODS IN MOLECULAR BIOLOGY

Series Editor
John M. Walker
School of Life and Medical Sciences
University of Hertfordshire
Hatfield, Hertfordshire, UK

For further volumes:
http://www.springer.com/series/7651

For over 35 years, biological scientists have come to rely on the research protocols and methodologies in the critically acclaimed *Methods in Molecular Biology* series. The series was the first to introduce the step-by-step protocols approach that has become the standard in all biomedical protocol publishing. Each protocol is provided in readily-reproducible step-by-step fashion, opening with an introductory overview, a list of the materials and reagents needed to complete the experiment, and followed by a detailed procedure that is supported with a helpful notes section offering tips and tricks of the trade as well as troubleshooting advice. These hallmark features were introduced by series editor Dr. John Walker and constitute the key ingredient in each and every volume of the *Methods in Molecular Biology* series. Tested and trusted, comprehensive and reliable, all protocols from the series are indexed in PubMed.

Retinal Gene Therapy

Methods and Protocols

Edited by

Livia S. Carvalho

The University of Melbourne, Parkville, VIC, Australia

Humana Press

Editor
Livia S. Carvalho
The University of Melbourne
Parkville, VIC, Australia

ISSN 1064-3745 ISSN 1940-6029 (electronic)
Methods in Molecular Biology
ISBN 978-1-0716-5197-1 ISBN 978-1-0716-5198-8 (eBook)
https://doi.org/10.1007/978-1-0716-5198-8

© The Editor(s) (if applicable) and The Author(s), under exclusive license to Springer Science+Business Media, LLC, part of Springer Nature 2026

This work is subject to copyright. All rights are solely and exclusively licensed by the Publisher, whether the whole or part of the material is concerned, specifically the rights of translation, reprinting, reuse of illustrations, recitation, broadcasting, reproduction on microfilms or in any other physical way, and transmission or information storage and retrieval, electronic adaptation, computer software, or by similar or dissimilar methodology now known or hereafter developed.
The use of general descriptive names, registered names, trademarks, service marks, etc. in this publication does not imply, even in the absence of a specific statement, that such names are exempt from the relevant protective laws and regulations and therefore free for general use.
The publisher, the authors and the editors are safe to assume that the advice and information in this book are believed to be true and accurate at the date of publication. Neither the publisher nor the authors or the editors give a warranty, expressed or implied, with respect to the material contained herein or for any errors or omissions that may have been made. The publisher remains neutral with regard to jurisdictional claims in published maps and institutional affiliations.

This Humana imprint is published by the registered company Springer Science+Business Media, LLC, part of Springer Nature.
The registered company address is: 1 New York Plaza, New York, NY 10004, U.S.A.

If disposing of this product, please recycle the paper.

Preface

The field of retinal gene therapy has witnessed remarkable progress in recent years, offering new hope for patients suffering from previously untreatable inherited retinal disorders. As we stand on the cusp of a new era in ophthalmology, this 2025 version of the book *Retinal Gene Therapy: Methods and Protocols*, aims to provide a comprehensive and up-to-date resource for researchers, clinicians, and students working in this rapidly evolving field.

The human retina, with its intricate network of neurons and supporting cells, presents both unique challenges and opportunities for gene therapy. Its accessibility and relatively immune-privileged status make it an ideal target for genetic interventions. However, the complexity of retinal diseases and the need for long-term efficacy and safety demand rigorous research and innovative approaches.

The path from laboratory to clinic for retinal gene therapies is fraught with regulatory complexities, reflecting the cutting-edge nature of these interventions. Researchers must navigate a labyrinth of requirements, including vector safety assessments, biodistribution studies, and long-term follow-up plans. Moreover, the rare nature of many inherited retinal diseases poses challenges in patient recruitment and trial design, often requiring innovative approaches such as adaptive trial designs or the use of natural history studies as comparators. Harmonizing regulatory requirements across different jurisdictions remains an ongoing challenge, particularly for multi-center, international trials. Despite these hurdles, collaborative efforts between researchers, clinicians, patient advocacy groups, and regulatory bodies are gradually streamlining the process, paving the way for more efficient translation of promising retinal gene therapies from bench to bedside.

This volume brings together contributions from leading experts in retinal gene therapy, covering a wide range of cutting-edge techniques and methodologies. The book is structured into four parts, each addressing critical aspects of the gene therapy development pipeline.

Part I focuses on vector development and design, exploring novel approaches to enhance the specificity and efficiency of gene delivery to retinal cells. The chapters in this section delve into the selection of retino-tropic AAV capsids through directed evolution and the application of CRISPR/CasRx technology for targeted gene knockdown.

Part II presents a series of in vitro assays crucial for validating retinal gene therapies. This section covers advanced techniques such as the generation of patient-specific iPSCs, the development of retinal organoids, and the use of retinal explants. These methods provide invaluable tools for modeling retinal diseases, screening potential therapies, and assessing their efficacy before moving to in vivo studies.

Part III shifts the focus to in vivo proof-of-concept studies and therapy validation in animal models. The protocols in this section guide researchers through various techniques, including intravitreal injections in mice, behavioral vision testing in dogs and zebrafish, and the design and validation of mirtron-mediated RNA interference for gene therapy.

Finally, Part IV addresses the critical step of clinical validation, presenting methods for assessing visual function in human subjects undergoing retinal gene therapy. These chapters provide insights into advanced perimetry techniques and comprehensive visual function assessments, essential for evaluating the efficacy of gene therapies in clinical trials.

Throughout the book, we have emphasized practical, step-by-step protocols that can be readily implemented in the laboratory. Each chapter includes detailed methods, tips for troubleshooting, and discussions of limitations and future perspectives. Our goal is to provide a valuable resource that will accelerate research and development in retinal gene therapy, ultimately leading to new treatments for patients with retinal disorders.

As the editor, I am deeply grateful to all the contributors who have shared their expertise and experience in this volume. Their dedication to advancing the field of retinal gene therapy is evident in the quality and depth of their contributions. I also extend my thanks to the team at Springer Nature for their support in bringing this book to fruition.

I hope that this book will serve as a catalyst for further innovation in retinal gene therapy, inspiring new research directions and fostering collaborations across disciplines. As we continue to unravel the complexities of retinal diseases and refine our therapeutic approaches, we move ever closer to the goal of restoring sight and improving the lives of millions affected by retinal disorders worldwide.

Parkville, VIC, Australia ***Livia S. Carvalho***

Contents

Contributors

EMILY L. ALLAN • *Lions Eye Institute, Nedlands, WA, Australia*

ALMAQDAD ALSALLOUM • *Moscow Institute of Physics and Technology, Dolgoprudny, Russia*

ANDREW J. ANDERSON • *Department of Optometry and Vision Sciences, Faculty of Medicine, Dentistry and Health Sciences, The University of Melbourne, Parkville, VIC, Australia*

DEBORAH AUBIN • *Stem Cell Medicine Group, School of Medical Sciences, Faculty of Medicine and Health, The University of Sydney, Sydney, NSW, Australia*

LAUREN N. AYTON • *Department of Optometry and Vision Sciences, Faculty of Medicine, Dentistry and Health Sciences, University of Melbourne, Parkville, VIC, Australia; Department of Surgery (Ophthalmology), Faculty of Medicine, Dentistry and Health Sciences, University of Melbourne, Parkville, VIC, Australia; Centre for Eye Research Australia, Royal Victorian Eye and Ear Hospital, Melbourne, VIC, Australia*

MICHAEL BACH • *Eye Center, Medical Center – University of Freiburg, Faculty of Medicine, University of Freiburg, Freiburg im Breisgau, Germany*

YASHVI BHATT • *Centre for Ophthalmology and Visual Science, The University of Western Australia, Crawley, WA, Australia; Retinal Genomics and Therapy Group, Lions Eye Institute Ltd, Nedlands, WA, Australia*

ALEXIS CEECEE BRITTEN-JONES • *Department of Optometry and Vision Sciences, Faculty of Medicine, Dentistry and Health Sciences, University of Melbourne, Parkville, VIC, Australia; Department of Surgery (Ophthalmology), Faculty of Medicine, Dentistry and Health Sciences, University of Melbourne, Parkville, VIC, Australia; Centre for Eye Research Australia, Royal Victorian Eye and Ear Hospital, Melbourne, VIC, Australia*

ALICIA A. BRUNET • *Centre for Ophthalmology and Visual Science, The University of Western Australia, Crawley, WA, Australia; Lions Eye Institute Ltd., Nedlands, WA, Australia; Faculty of Medicine, Dentistry and Health Sciences, Department of Optometry and Vision Sciences, University of Melbourne, Parkville, VIC, Australia*

BANG V. BUI • *Department of Optometry and Vision Sciences, University of Melbourne, Parkville, VIC, Australia*

LIVIA S. CARVALHO • *Centre for Ophthalmology and Visual Science, The University of Western Australia, Crawley, WA, Australia; Retinal Genomics and Therapy Group, Lions Eye Institute Ltd, Nedlands, WA, Australia; Faculty of Medicine, Dentistry and Health Sciences, Department of Optometry and Vision Sciences, University of Melbourne, Parkville, Australia*

SHANE CHEUNG • *Biological Optical Microscopy Platform, The University of Melbourne, Melbourne, VIC, Australia*

ELLIE CHO • *Biological Optical Microscopy Platform, The University of Melbourne, Melbourne, VIC, Australia*

JANISE DAGELET • *Department of Optometry and Vision Sciences, Faculty of Medicine, Dentistry and Health Sciences, University of Melbourne, Parkville, VIC, Australia; Department of Surgery (Ophthalmology), Faculty of Medicine, Dentistry and Health Sciences, University of Melbourne, Parkville, VIC, Australia; Centre for Eye Research Australia, Royal Victorian Eye and Ear Hospital, Melbourne, VIC, Australia*

NATHAN EHRENREICH • *Department of Ophthalmology, Ocular Genomics Institute, Massachusetts Eye and Ear Infirmary, Harvard Medical School, Boston, MA, USA*

Paula I. Fuller-Carter • *Lions Eye Institute Ltd., Nedlands, WA, Australia*
Marcela Garita-Hernandez • *Department of Ophthalmology, Ocular Genomics Institute, Massachusetts Eye and Ear Infirmary, Harvard Medical School, Boston, MA, USA*
Anai Gonzalez-Cordero • *Stem Cell Medicine Group, School of Medical Sciences, Faculty of Medicine and Health, The University of Sydney, Sydney, NSW, Australia*
Jenna Hall • *Department of Anatomy and Physiology, The University of Melbourne, Melbourne, VIC, Australia*
Susan Haller • *Department of Research Pathology, Genentech, Inc., South San Francisco, CA, USA*
Ruth E. Hogg • *School of Medicine, Dentistry and Biomedical Sciences, Centre for Public Health, Queen's University, Belfast, UK*
Madina Khubetsova • *S.N. Fedorov National Ophthalmology Medical Research Center "Eye Microsurgery", Moscow, Russia*
Daria Kolotova • *Institute of Higher Nervous Activity and Neurophysiology of Russian Academy of Sciences, Moscow, Russia*
Satheesh Kumar • *Centre for Eye Research Australia, Royal Victorian Eye and Ear Hospital, East Melbourne, VIC, Australia; Ophthalmology, Department of Surgery, University of Melbourne, East Melbourne, VIC, Australia; Menzies Institute for Medical Research, University of Tasmania, Hobart, TAS, Australia*
Grace E. Lidgerwood • *Department of Anatomy and Physiology, The University of Melbourne, Melbourne, VIC, Australia*
Leszek Lisowski • *Translational Vectorology Research Unit, Children's Medical Research Institute, Faculty of Medicine and Health, The University of Sydney, Westmead, NSW, Australia; Australian Genome Therapeutics Centre, Children's Medical Research Institute and Sydney Children's Hospitals Network, Westmead, NSW, Australia; Laboratory of Molecular Oncology and Innovative Therapies, Military Institute of Medicine - National Research Institute, Warszawa, Poland*
Guei-Sheung Liu • *Centre for Eye Research Australia, Royal Victorian Eye and Ear Hospital, East Melbourne, VIC, Australia; Ophthalmology, Department of Surgery, University of Melbourne, East Melbourne, VIC, Australia; Menzies Institute for Medical Research, University of Tasmania, Hobart, TAS, Australia*
Robert E. MacLaren • *Nuffield Department of Clinical Neurosciences & NIHR Oxford Biomedical Research Centre, University of Oxford, UK and Oxford Eye Hospital, Oxford University Hospitals NHS Foundation Trust, Oxford, UK*
Alexey Malyshev • *Institute of Higher Nervous Activity and Neurophysiology of Russian Academy of Sciences, Moscow, Russia*
Rebecca M. Marton • *Department of Regenerative Medicine, Genentech, Inc., South San Francisco, CA, USA*
Michelle E. McClements • *Nuffield Department of Clinical Neurosciences & NIHR Oxford Biomedical Research Centre, University of Oxford, UK and Oxford Eye Hospital, Oxford University Hospitals NHS Foundation Trust, Oxford, UK*
Thibaud Metais • *Department of Ophthalmology, Ocular Genomics Institute, Massachusetts Eye and Ear Infirmary, Harvard Medical School, Boston, MA, USA*
Stylianos Michalakis • *Department of Ophthalmology, LMU University Hospital, LMU Munich, Munich, Germany*
Annie L. Miller • *Centre for Ophthalmology and Visual Science, The University of Western Australia, Crawley, WA, Australia; Retinal Genomics and Therapy Group, Lions Eye Institute Ltd, Nedlands, WA, Australia*

Olga Mityaeva • *Moscow Institute of Physics and Technology, Dolgoprudny, Russia*
Jessica K. Mountford • *Lions Eye Institute, Nedlands, WA, Australia; Centre for Ophthalmology and Visual Sciences, The University of Western Australia, Perth, WA, Australia; School of Life Sciences, La Trobe University, Melbourne, VIC, Australia*
Freya M. Mowat • *Department of Surgical Sciences, University of Wisconsin-Madison School of Veterinary Medicine, Madison, WI, USA; Department of Ophthalmology and Visual Sciences, University of Wisconsin-Madison School of Medicine and Public Health, Madison, WI, USA*
Harry O. Orlans • *Nuffield Department of Clinical Neurosciences & NIHR Oxford Biomedical Research Centre, University of Oxford, UK and Oxford Eye Hospital, Oxford University Hospitals NHS Foundation Trust, Oxford, UK*
Daniel Paull • *The New York Stem Cell Foundation Research Institute, New York, NY, USA*
Marina Pavlou • *Department of Neurobiology and Biophysics, University of Washington, Seattle, WA, USA*
Alice Pébay • *Department of Anatomy and Physiology, The University of Melbourne, Melbourne, VIC, Australia; Department of Surgery, Royal Melbourne Hospital, The University of Melbourne, Melbourne, VIC, Australia*
Simon M. Petersen-Jones • *Department of Small Animal Clinical Sciences, Michigan State University College of Veterinary Medicine, East Lansing, MI, USA*
Eliah R. Shamir • *Department of Research Pathology, Genentech, Inc., South San Francisco, CA, USA*
Kalyan Shobhana • *Biological Optical Microscopy Platform, The University of Melbourne, Melbourne, VIC, Australia*
Federica Staurenghi • *Nuffield Department of Clinical Neurosciences & NIHR Oxford Biomedical Research Centre, University of Oxford, UK and Oxford Eye Hospital, Oxford University Hospitals NHS Foundation Trust, Oxford, UK*
Pavel Volchkov • *Moscow Institute of Physics and Technology, Dolgoprudny, Russia*
Adrian Westhaus • *Translational Vectorology Research Unit, Children's Medical Research Institute, Faculty of Medicine and Health, The University of Sydney, Westmead, NSW, Australia; INSERM U951, Genethon, Évry-Courcoronnes, France*
Vickie H. Y. Wong • *Department of Optometry and Vision Sciences, University of Melbourne, Parkville, VIC, Australia*
Da Zhao • *Department of Optometry and Vision Sciences, University of Melbourne, Parkville, VIC, Australia*
Qingyun Zheng • *Nuffield Department of Clinical Neurosciences & NIHR Oxford Biomedical Research Centre, University of Oxford, UK and Oxford Eye Hospital, Oxford University Hospitals NHS Foundation Trust, Oxford, UK*

Part I

Retinal Gene Therapy Vector Development and Design

Chapter 1

Selecting Novel Retino-tropic AAV Capsids by Directed Evolution

Adrian Westhaus and Leszek Lisowski

Abstract

The selection of novel AAV variants for retinal transduction has been successful for several model systems in the past. Despite this, the publication of a full directed evolution protocol that explains all steps from basic library generation techniques to the selection of suitable candidates for further testing is lacking. The following methods chapter provides a detailed protocol for the generation of an AAV library and the recovery of the most functional candidates using highly predictive transgene expression-driven selection platform. It is important to note that multiple selection platforms have been developed to date, each with unique set of strengths and weaknesses. However, the choice of a selection platform most suitable for given AAV bioengineering application is beyond the scope of this chapter. For the purpose of this chapter, we will describe steps involved in generation of an exemplar peptide insertion library into the variable region VIII of the AAV2 capsid, but the steps outlined can be extended to all other serotypes and other surface loops. Furthermore, the technologies described work independently of the model system and do not require any major specialist equipment.

Key words AAV, Directed evolution, Capsid library selection, Gene therapy, Viral vector production

1 Introduction

Many genetic defects can impair vision and ultimately lead to blindness; therefore, many clinical trials, including those using adeno-associated virus (AAV)-derived vectors, have been undertaken to treat these defects [1]. Some of the clinical programs have led to exciting successes, with the most prominent example being studies that formed the foundation of a market-approved drug LUXTURNA (voretigene neparvovec-rzyl) [2, 3]. As new preclinical model systems are developed, and further advances in AAV capsid bioengineering are achieved, it is important to apply the new available tools to develop novel capsid variants. These next-generation AAV variants will enable more efficient transduction of the target cells in the human retina, supporting the development of safer and more effective novel advanced therapies.

Livia S. Carvalho (ed.), *Retinal Gene Therapy: Methods and Protocols*, Methods in Molecular Biology, vol. 3023,
https://doi.org/10.1007/978-1-0716-5198-8_1,

© The Author(s), under exclusive license to Springer Science+Business Media, LLC, part of Springer Nature 2026

Directed evolution-based selection of novel AAVs from large "random" mixes of AAV variants (known as AAV capsid libraries) for retinal transduction has been successful in rodents and nonhuman primate models [4–6]. Despite this, the publication of a detailed directed evolution protocol that outlines all the steps from basic library generation techniques to the selection of suitable candidates for further testing is presently lacking.

This protocol describes the application of the functional transduction (FT) platform, which enables the selection of novel capsid variants based on transgene expression [7], for screening of a bespoke peptide display library using any model of the human retina.

2 Materials

Of note: many of the items listed can be substituted by alternative similar products and/or may become obsolete with advancing technologies. The following lists contain materials that were rigorously validated to yield robust results.

2.1 Specialist Equipment

1. Ultracentrifuge: Beckman Coulter Optima XPN or equivalent ultracentrifuge.
2. PCR system with heated lid, e.g., Mastercycler Pro Thermal Cycler.
3. Electroporator: Bio-Rad MicroPulser Electroporator or equivalent.
4. Swinging Bucket Rotor SW32Ti or equivalent.
5. Swinging Bucket Rotor SW41Ti or equivalent.
6. Centrifuge: Eppendorf 5425 R or equivalent centrifuge.

2.2 Molecular Biology Processing Reagents and Consumables

1. 0.2 mL 8-strip PCR tubes.
2. 1 kb plus DNA ladder.
3. 15 mL tubes.
4. 50 mL tubes.
5. 50× TAE Buffer DNA Typing Grade: 2 M Tris-acetate, 50 mM EDTA. Filter sterilized.
6. 500 mL centrifuge tubes.
7. Agarose.
8. Agar bacteriological.
9. AllPrep DNA/RNA Kit.
10. β-Mercaptoethanol.
11. Deoxynucleotide (dNTP) Solution Mix.

12. DH-5-alpha bacteria.
13. 100% ethanol, absolute.
14. Electroporation cuvette.
15. Gel loading dye.
16. Klenow fragment (3′ → 5′ exo-).
17. NEBuilder HiFi DNA assembly.
18. NsiI restriction enzyme.
19. Nuclease-free water.
20. Endotoxin-Free Maxi Kit.
21. High-fidelity DNA polymerase.
22. Miniprep Kit.
23. PCR purification kit.
24. RNAse away surface decontaminant.
25. Safe-lock tubes 0.5 mL.
26. Safe-lock tubes 1.5 mL.
27. SfiI restriction enzyme.
28. 3 M sodium acetate (pH 5.2).
29. SS320 electrocompetent cells or equivalent.
30. cDNA Synthesis Kit such as SuperScript IV First-Strand Synthesis System.
31. SwaI restriction enzyme.
32. T4 DNA ligase.
33. Trimethoprim.
34. Tubes 5 mL PCR clean.
35. DNase inactivation reagent.
36. Water for irrigation.
37. Yeast extract.
38. Gel DNA Recovery Kit.

2.3 Vector Production and Tissue Culture Reagents and Consumables

1. 100 kDa Amicon Ultra-15 or equivalent.
2. 250 U/uL Benzonase.
3. Costar Spin-X, 0.22 μm filter, centrifuge tube or equivalent.
4. Human embryonic kidney (HEK) 293 T cells.
5. OptiMEM media, 500 mL.
6. PEG 8000.
7. PEI (polyethylenimine MW 40,000).
8. 10% Pluronic F-68 solution.
9. Sodium deoxycholate (DOC).

10. Animal origin-free recombinant enzyme for dissociating adherent mammalian cells.
11. 32 mL ultracentrifuge tubes (first CsCl spin) such as from Beckman Coulter.
12. 12 mL ultracentrifuge tubes (second CsCl spin) such as from Beckman Coulter.
13. Virkon Tablets.

2.4 Cell Culture Media for Bacterial Culture and Vector Production

1. LB media: 1 L de-ionized water, 10 g tryptone, 10 g sodium chloride, 5 g yeast extract.
2. LB agar: 250 mL LB media, 3.75 g agar bacteriological.
3. HEK293T media for vector production: 500 mL DMEM, 50 mL FBS, 5 mL penicillin–streptomycin, 12.5 mL 1 M HEPES.

2.5 Reagents for Vector Production

1. Cell pellet resuspension buffer: 496 mL DPBS, 5 mL 1 M Tris–HCl pH 8.5, 1 mL 1 M $MgCl_2$.
2. PEG-media pellet resuspension buffer: 496 mL DPBS, 5 mL 1 M Tris–HCl pH 8.0, 1 mL 1 M $MgCl_2$.
3. Na-HEPES-EDTA solution: 435 mL pyrogen-free water for irrigation, 25 mL 1 M HEPES, 25 mL 0.5 M EDTA, 15 mL 5 M sodium chloride.
4. 1.30 g/cm^3 CsCl: 39.9 g CsCl, 89.9 mL DPBS (refractive index [RI]: 1.3619).
5. 1.372 g/cm^3 CsCl: 49.7 g CsCl, 87.3 mL DPBS (RI: 1.3689).
6. 1.50 g/cm^3 CsCl: 74.4 g CsCl, 80.6 mL DPBS (RI: 1.3807).

3 Methods

Library Generation at a Glance The AAV capsid gene can incorporate insertions of short peptides in VR-VIII. The library plasmid backbone is generated by providing appropriate acceptor sites in VR-VIII and transfer of the modified capsid gene into a plasmid containing AAV inverted terminal repeats (ITRs, Fig. 1a). The acceptor sites are then used for large-scale cloning of randomized short peptides (usually seven to ten amino acids in length), generating a peptide insertion library (Fig. 1b). The newly generated library is mixed with helper plasmids and transfected into HEK293T cells using standard methods. Following harvest and purification using iodixanol or CsCl gradient centrifugation, the packaged library can be used as input for directed evolution experiments. Additionally, the cloned plasmid library and the purified vector library should be used as PCR templates for amplification of the VR-VIII target region and submission to deep sequencing. This verifies the diversity and quality of the input library.

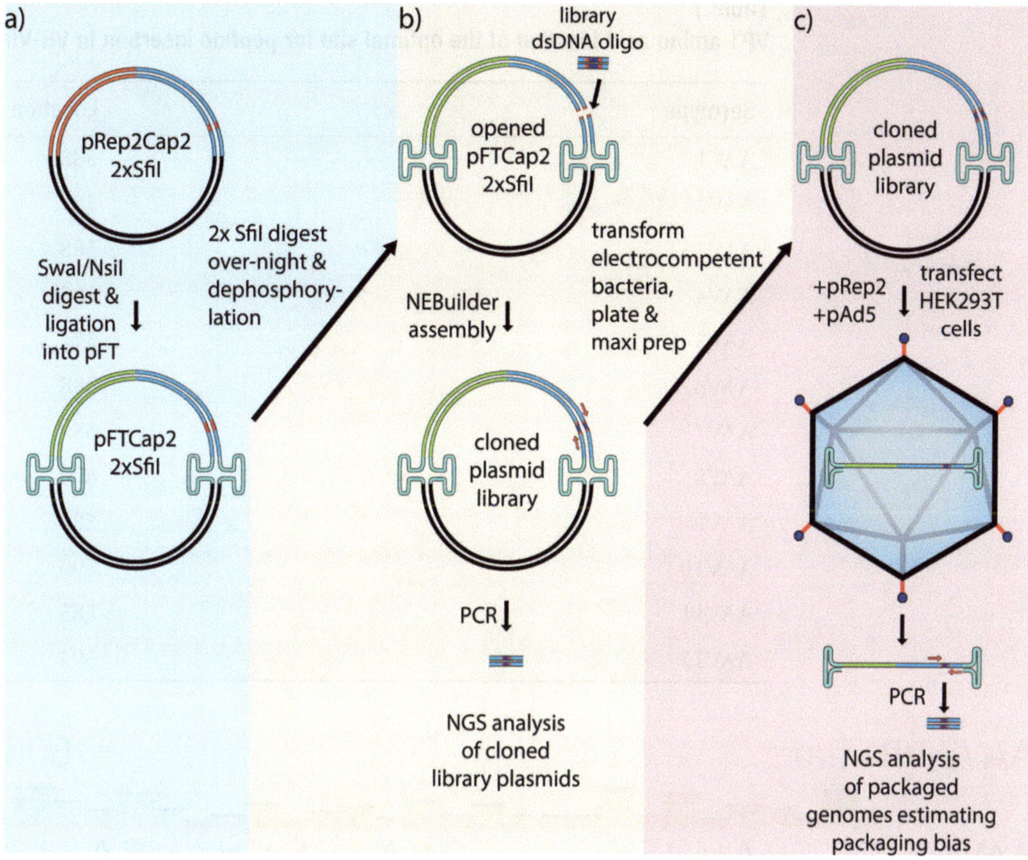

Fig. 1 Library generation workflow at a glance. (**a**) Preparation of the library plasmid containing the pFT platform and the AAV2 capsid gene with the double SfiI acceptor sites. (**b**) Cloning of the double-stranded randomized library oligo into the SfiI-digested library plasmid and transformation into electrocompetent bacteria. (**c**) Finalized plasmid library which will be co-transfected into human embryonic kidney (HEK) 293 T cells to produce the final packaged library

Preparing the Library Plasmid There are several methods of introducing a peptide insertion into VR-VIII of various AAV serotypes of choice. The method herein described is based on the use of double SfiI restriction sites first described by Müller and colleagues [8]. The exact site of insertions (first described for AAV2) was further refined and extended to many other AAV variants [9]. Utilizing these findings, we were able to predict the corresponding insertion site for many other serotypes, including AAV10 and AAV11 (Table 1).

Generally speaking, the insertion site for any mammalian-derived AAV capsid can be predicted by multiple sequence alignment. The insertion site is always three amino acids downstream of a highly conserved glutamine (Q584 based on AAV2 VP1 numbering, Fig. 2). Using these predictions, we have generated a pRep2Cap2-SfiI as described in a previous publication [7]. The

Table 1
VP1 amino acid location of the optimal site for peptide insertion in VR-VIII

Serotype	Location
AAV1	588
AAV2	587
AAV3b	588
AAV4	585
AAV5	577
AAV6	588
AAV7	589
AAV8	890
AAV9	588
AAV10	590
AAV11	585
AAV12	594

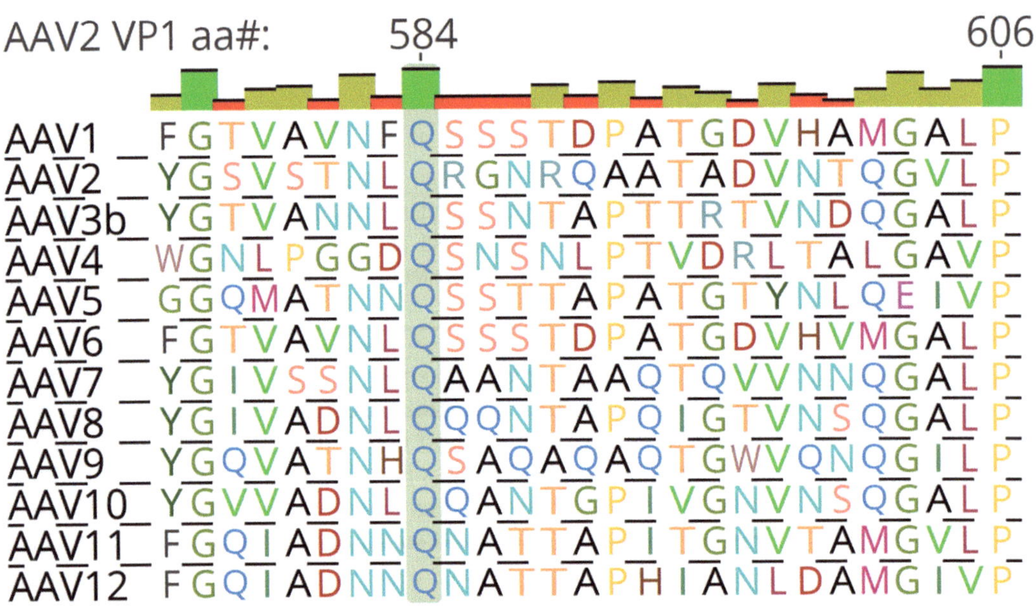

Fig. 2 Alignment of variable region VIII of AAV1 to AAV12. Highlighted in green is a highly conserved glutamine, Q584, based on AAV2 VP1 numbering. The principal insertion site for peptide insertion libraries in this surface loop is always three amino acids downstream of this glutamine residue

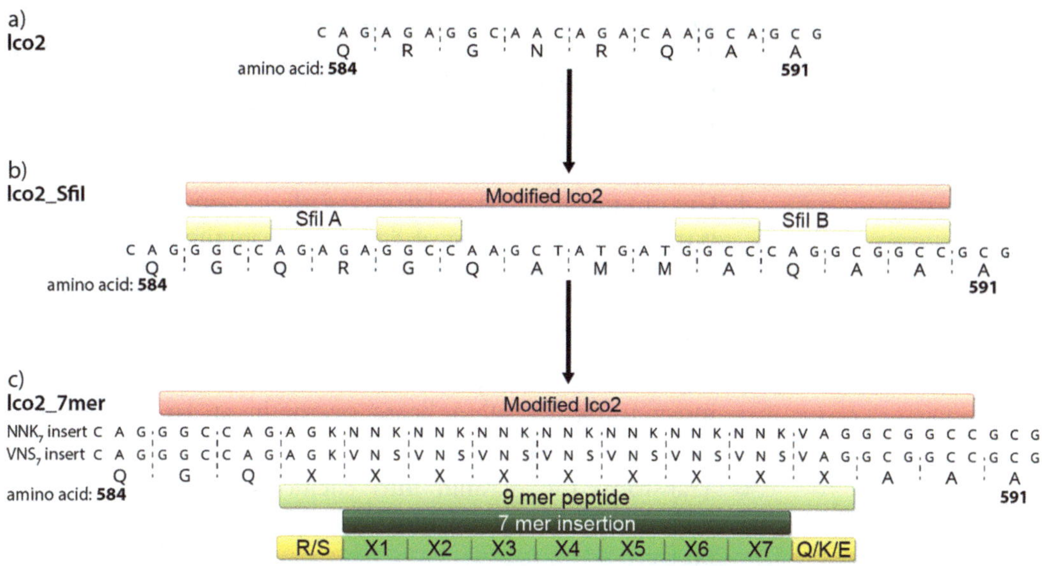

Fig. 3 Modifications in the VR-VIII of AAV2 and library cloning. Detailed view of the modification workflow of the local codon optimized (lco)2 capsids. (**a**) Unmodified sequence of lco2. The highly conserved Q584 is highlighted throughout. (**b**) Insertion of two SfiI restriction sites. (**c**) Insertion of the peptide library. Seven truly randomized NNK or VNS (X1-X7) insertions as well as the full 9mer peptide including semi-random flanking amino acids are shown within the modified region. (This figure is modified from Westhaus et al. (2022) [7])

same modification was performed for AAV1 to AAV12, in the context of local codon optimized version of the *cap* gene [10].

As the library generation itself is almost identical for all parental AAV serotypes, this protocol will focus on the generation of a peptide insertion library in the local codon optimized version of AAV2. In summary, the local codon optimized AAV2 capsid was modified to incorporate double SfiI restriction sites, replacing six amino acids downstream of Q584 (Fig. 3a, b). In the pRep2Cap2-SfiI plasmids, the cap gene is flanked by SwaI and NsiI restriction sites which were used to transfer the modified capsid containing the double SfiI restriction sites into the FT selection platform (Fig. 1a). The double SfiI modified *cap2* gene, the 11 other capsid genes harboring homologous modifications (Table 1), and the plasmid encoding the FT selection platform can be obtained from the Lisowski lab under an MTA or licensing agreements.

Detailed Steps for Library Plasmid Backbone Preparation:

1. Grow bacterial cultures of pRep2Cap2-SfiI (ampicillin-resistance) and pFT (trimethoprim-resistance) plasmids and purify the plasmid with PureLink Expi Endotoxin-Free Maxi Kit or similar plasmid purification kit.

2. In separate 50 μL reactions, digest 5 μg of the plasmids with 1.5 μL each SwaI and NsiI (not the HF version; *see* **Note 1**) in buffer 3.1 for 8 h at 25 °C followed by 3 h at 37 °C.
3. Prepare a 1% agarose gel and run both digests until the bands are well separated.
4. Cut out the ~2.3 kb band from the pRep2Cap2-SfiI digest (this is the *cap* gene) and the ~4.8 kb band from the pFT digest (this is the library backbone without *cap* gene) and purify digested linear DNA with Zymoclean Gel DNA Recovery or similar gel extraction kit.
5. Ligate 60 ng of the pFT backbone with 86 ng of the *cap* gene insert using T4 DNA ligase overnight at 16 °C following manufacturer's protocol.
6. Transform the plasmid into chemo-competent DH5-alpha K12 *Escherichia coli* bacteria with the ligation mix for 30 min on ice.
7. Heatshock the incubation mix for 60 s at 42 °C, recover on ice for 1 min, and add 200 μL of 37 °C LB media.
8. Recover bacteria in shaker at 37 °C for 30 min.
9. Plate 100 μL of bacterial solution on LB-agar plates with trimethoprim (10 μg/mL final concentration). Incubate at 37 °C overnight.
10. Inoculate 5 mini-prep cultures with individual bacterial colonies and incubate at 37°C overnight in bacterial shaker.
11. Perform mini-prep with a suitable kit (QIAPrep Spin Miniprep Kit or similar) according to manufacturers' instructions.
12. Check for the correct cloning product by diagnostic digest (SfiI, 50 °C for 2 h) and sanger sequencing (SeqCapF [CACTGCCTGCGATCTGGTC] and SeqCapR [CCGCCATGCTACTTATCTACATGC]).
13. Maxi-prep correct clone following the manufacturers' instructions.
14. Use 10 μg of purified plasmid and perform SfiI digest at 50 °C overnight. *See* **Note 2**.
15. Purify the digested plasmid with a PCR purification kit or magnetic beads (QIAGEN or AMPure) and set up another SfiI digest at 50 °C overnight. This is to ensure that all the plasmid is digested, which removed background self-ligation of the backbone.
16. Add 1 μL of calf-intestinal phosphatase (e.g., Quick CIP, NEB) to each 50 μL digest and incubate at 37 °C for 15 min. Inactivate by adding EDTA and SDS containing loading dye.

17. Prepare a 1% agarose gel and run the digest until the single ~7.1 kb band is well visible.
18. Cut the band from the gel and purify as described above (**step 4**).

Cloning the Plasmid Library Introducing the randomized peptide into VR-VIII can be performed in multiple ways. Herein, we will describe a method that introduces a hepta-repeat of NNK codons (NNK_7) and a hepta-repeat of VNS codons (VNS_7) by utilizing an updated version of Gibson assembly, called NEBuilder HiFi assembly. The reason for creating both types of library inserts is that the NNK_7 library yields fully randomized libraries but can also introduce the UAG stop codon, which will prevent expression of the full VP sequence. The VNS_7 library on the other hand avoids all UNN codons and therefore will never have a stop codon. Other unfavorable amino acids such as cysteine are also avoided, but the overall diversity will be lower. A mix of the two libraries will yield a high diversity, highly functional final library (Fig. 3c).

Detailed Steps for Library Plasmid Generation

1. Obtain the following cartridge-purified oligonucleotides (*see* **Note 3**):

Capsid	Oligonucleotides (library and complementary)
AAV2	accaacctccagggccagAG**KNNKNNKNNKNNKNNKNNKNNKVA**Ggcggccgcgaccgcagat
	accaacctccagggccagAG**KVNSVNSVNSVNSVNSVNSVNSVA**Ggcggccgcgaccgcagat
	ATCTGCGGTCGCGGCCGC

2. Prepare a second-strand synthesis reaction using the polymerase activity of the Klenow (exo-) fragment. In a 50 μL reaction mix.
3. 5 μL NEB buffer 2
   (a) 2 μL long library containing oligo (100 μM), only one of the two per reaction.
   (b) 2 μL short complementary oligo.
   (c) 1 μL dNTPs.
   (d) In a thermocycler, heat the mix to 98 °C and let the reaction cool down to 50 °C as slowly as possible (usually at −0.1 °C per minute) to anneal the oligos.
4. Once the reaction reaches 50 °C, add 3 μL of Klenow (exo-), mix with P200 pipette, change the temperature to 37 °C, and allow the second-strand synthesis to proceed for 30 min.

5. Prepare a 3% agarose gel and run the reaction mix until the <100 bp bands are well visible. Gel purify as described above (**step 4**). The resulting products are the NNK_7 library dsDNA oligo and the VNS_7 library dsDNA oligo.
6. Set up a large-scale assembly reaction separately for the NNK_7 and VNS_7 libraries by mixing:
   (a) 987 ng (225 fmol) Sfi-digested pFT-Cap2-SfiI.
   (b) 87.7 ng (2250 fmol) dsDNA library oligo.
   (c) Fill up to 80 μL with nuclease-free water.
   (d) Add 80 μL NEBuilder HiFi DNA assembly and mix well.
   (e) Separate into eight PCR tubes and incubate at 50 °C for 30 min.
7. Desalting: Combine all eight reactions into one 1.5 mL tube and add 16 μL 3 M sodium acetate and 400 μL 100% ethanol, analytical grade. Incubate mix at −80 °C for 30 min (*see* **Note 4**).
8. Spin for 15 min at 4 °C and > 15,000 ×g. Carefully discard all supernatant.
9. Add 1000 μL ice-cold 70% ethanol and spin for 15 min at 4 °C and >15,000 ×g. Carefully discard the supernatant completely. Repeat one more time.
10. Let the pellet dry for 5 min at room temperature and resuspend in 20 μL nuclease-free water by incubating for 10 min at 56 °C.
11. Thaw SS320 bacterial cells on ice. Cool down electroporation cuvette and DNA on ice. Warm SS320 recovery media (provided with SS320 bacteria) to 37°C.
12. Prepare clean bench with Bunsen burner running.
13. Mix bacterial cells with 20 μL assembled and desalted DNA on ice and transfer to ice-cold electroporation cuvette.
14. Dry electroporation cuvette and place into the Bio-Rad Micro-Pulser. Select program EC1 for 1 mm cuvette or EC2 for 2 mm cuvette. Run the electroporation.
15. After the pulse, immediately add 1 mL warm recovery media to electroporated bacteria in the cuvette. Use the media to transfer the entire reaction into 5 mL inoculation tube. Use another 1 mL warm recovery media to carefully rinse the electroporation cuvette and transfer liquid to the same 5 mL tube.
16. Place tube into 37 °C bacterial shaker for 30 min.
17. Prepare a dilution series from 10^{-1} to 10^{-5} of the recovered bacteria in LB media and plate 50 μL of each dilution onto 10 cm LB-agar plate with 10 μg/mL trimethoprim. Incubate at 37°C overnight. Count colonies the next day to estimate

transformation efficiency (*see*, e.g., https://www.nebiolabs.com.au/faqs/0001/01/01/how-should-i-calculate-the-transformation-efficiency-c2992). Based on our experience, cloning that yields more than 10^7 clones is acceptable (*see* **Note 5**).

18. In parallel to **step 17**, transfer remaining culture to 100 mL LB media with 10 μg/mL trimethoprim in a beveled shaker flask. Place into 37 °C bacterial shaker overnight (*see* **Note 5**).
19. The following day, assuming the quality control (**step 17**) returns a positive result, maxi-prep plasmid library using an endotoxin-free maxiprep kit to generate transfection-grade plasmid library.

Viral Vector Manufacturing AAV production is performed in adherent HEK293T cells grown on 15 cm tissue culture-treated dishes in DMEM supplemented with 10% fetal bovine serum (FBS), 1% Pen–Strep, and 25 mM HEPES. Transfection is performed with linear polyethyleneimine hydrochloride (PEI MAX) at 2:1 PEI/DNA ratio. The transfection mix contains three plasmids: (1) 11.6 kb pAd5 plasmid encoding adenoviral helper proteins *E2A*, *E4*, and *VA*, (2) 5.2 kb pRep2 encoding replication helper genes (*rep2*) of AAV2 (pRep2 is used for selections using the following platforms: CREATE [11], iTransduce [12], TRACER [13], TRADE [14], and FT [7]), and (3) 7.1 kb library plasmid containing AAV ITRs (pFT-Cap2-NNK_7/VNS_7, in this protocol). The plasmids are mixed at a molar ratio of 2:1:0.01 (pAd5/pRep2/pFT, Fig. 4). The suggested ratio of individual plasmids was validated to reduce the amount of cross-packaging [7, 15]. The resulting packaged library can be processed and purified in several different ways. This protocol describes cesium chloride (CsCl) gradient centrifugation, as this is the least biased purification method. The purification is based on density rather than capsid sequence/surface. Therefore, no biases are being introduced, and CsCl gradients still enable a relatively large-scale manufacturing and high purity.

Detailed Steps for the Manufacturing of a Viral Vector Library as a 60 × 15 cm Dish Prep
See **Note** 7.

1. Split HEK293T cells into 60× 15 cm plates using 25 mL media per 15 cm dish. Seed ~11 to 13 × 10^6 cells/dish for ~70% confluency the following day.
2. Incubate at 37 °C, 5% CO_2, in humidified atmosphere overnight.
3. Carefully replace media with 15 mL fresh warm complete media 2 h prior to transfection.
4. Measure the concentrations of all necessary plasmids.

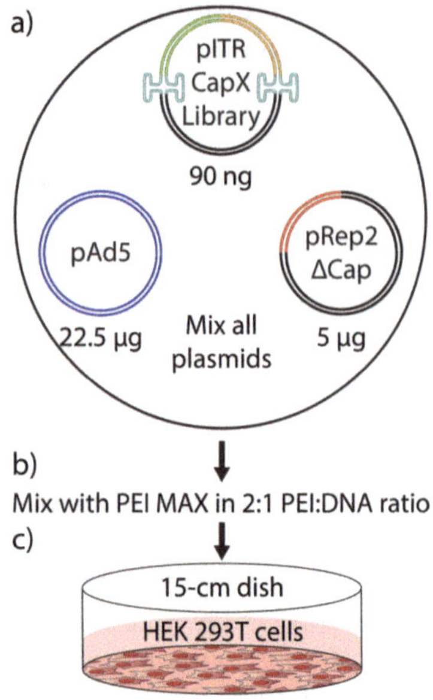

Fig. 4 Schematic of producer cell transfection. (**a**) Prepare a plasmid mix following a molar ratio of 2:1:0.01 (pAd5/pRep2/pFT). (**b**) Mix with PEI MAX in OptiMEM. (**c**) Add to HEK293T cells

5. Prepare a mix of 3.7 mL PEI MAX (1 mg/mL in nuclease-free water, pH 7) and 13.6 mL OptiMEM.
6. Mix the following plasmids:
   (a) 5.94 µg pFT-Cap2-library (either a 1:1 mix of NNK_7 and VNS_7 libraries or both libraries are manufactured separately).
   (b) 330 µg pRep2.
   (c) 1485 µg pAd5.
   (d) Fill to 16.5 mL with OptiMEM.
7. Add 16.5 mL PEI-OptiMEM from **step 42**.
8. Mix gently but thoroughly and incubate for 10 min at ambient temperature.
9. Add 500 µL to each 15 cm dish and gently swirl the dish to distribute the transfection mix.
10. Incubate at 37 °C, 5% CO_2, in humidified atmosphere for 72 h.
11. Harvest cells and media into 500 mL centrifuge tubes by scraping or pipetting up and down.
12. Spin 5 min at maximum speed in tabletop centrifuge (5250 × g, 4 °C).
13. Transfer supernatant to fresh tubes, add 1/4 volume 40% PEG 8000/2.5 M NaCl (final 8%), and incubate for 3–12 h at 4 °C.

14. Resuspend cell pellet from **step 49** in 20 mL cell pellet resuspension buffer (Table 1), transfer to 50 mL tube, and break cells with three freeze/thaw cycles. Keep at −80 °C until media from **step 50** is processed.
15. Spin PEG-precipitated media (from **step 50**) 30 min at maximum speed (5250 × g, 4 °C).
16. Resuspend media pellet in 20 mL PEG-media pellet resuspension buffer (Table 1).
17. Combine both resuspended pellets (from **steps 51** and **53**) and add 36 μL Benzonase (200 U/mL final concentration). Incubate at 37 °C for 60 min.
18. Add 1/4 volume 5 M NaCl (final 1 M) and 10% DOC to a final concentration of 0.5%. Incubate at 37 °C for 30 min.
19. Spin for 30 min at maximum speed (5250 × g, 4 °C).
20. Transfer supernatant to new 50 mL tubes and add 1/4 volume 40% PEG 8000/NaCl (final 8%) and incubate for 3–12 h at 4 °C.
21. Spin for 30 min at maximum speed (5250 × g, 4 °C) and discard supernatant.
22. Resuspend in 40 mL Na-HEPES-EDTA (Table 1) and keep at 4 °C.
23. Prepare the following CsCl solutions. Verify using refractometer:

    CsCl in PBS:

Density (g/cm^3)	CsCl (g)	PBS (mL)	RI
1.30	39.9	89.9	1.3619
1.372	49.7	87.3	1.3689
1.50	74.4	80.6	1.3807

24. Prepare two 32 mL UltraClear centrifuge tubes for each spin.
25. Add 12 mL 1.3 g/cm^3 to each tube.
26. Carefully underlay 5 mL 1.5 g/cm^3 underneath the 1.3 g/cm^3 layer. Mark interface.
27. Add 20 mL of resuspended viral vector solution to the top of the gradient.
28. Balance tubes and spin in ultracentrifuge using SW32 Ti rotor at 25 K RPM (106,800 ×g) at 20 °C for ~24 h (minimum 20 h).
29. Extract 5 mL of the solution after the spin by inserting an 18 gauge needle into the spot that marked the 1.3/1.5 g/cm^3 interface.

30. Mix the 5 mL extracted volume from both spins and measure the refractive index. Adjust the refractive index to 1.3689 by using the 1.3/1.5 g/cm^3 solutions. Fill to 12 mL with the 1.372 g/cm^3 solution and transfer to a 13.2 mL UltraClear centrifuge tube.
31. Prepare a balance tube with 1.372 g/ cm^3 solution. Adjust weight of both tubes to the nearest 0.01 g.
32. Spin in ultracentrifuge SW41 Ti rotor at 38 K RPM (247,600 × g) at 20 °C for 24 h.
33. Collect visible vector band in under 3 mL total volume using 5 mL syringe/18 G needle (*see* **Note 9**).
34. Dialyze the vector prep in a 3 mL Slide-A-Lyzer dialysis cassette in 2 L PBS for 6 h.
35. Repeat **step 71**.
36. Concentrate dialyzed vectors in a 100 kDa concentrator to obtain volumes between 250 μL and 500 μL.
37. Sterilize vector prep by adding into a Costar Spin-X 0.22 μm filter column and spin at 10,000 × g for 5 min. Transfer sterile prep to cryovial.
38. Determine the genomic titer of the library using quantitative PCR protocols such as qPCR, dPCR, or ddPCR, following the manufacturer's instructions. The following primers can be used for SYBR/EVA green quantification of an FT platform library:

Primer pair	Name	Sequence
eGFP	eGFP_fwd	GAGGTGAAGTTCGAGGGC
	eGFP_rev	CTTGTGCCCCAGGATGTTG
Rep2	Rep2_fwd	CTCAACCCGTTTCTGTCGTC
	Rep2_rev	CACATTGACCAGATCGCAGG

Transduction of a Relevant Model of Retinal Cells and Recovery of Successful Variants

Directed evolution is performed as iterative rounds of enrichment of high-performing candidates. Starting with an initial highly diverse library (Fig. 5a), target cells are transduced at range of doses (Fig. 5b), and, after an appropriate duration of exposure to AAV, DNA and RNA are extracted from them. The AAV capsid DNA would have been introduced to the cells by AAV capsids that entered the cells well, while the AAV capsid-encoding RNA would have been transcribed from successful capsids that entered the nucleus and enabled gene expression. As the FT platform enables the expression of the capsid gene in a variety of target cells, the RNA is usually used as a template for variant recovery (Fig. 5c).

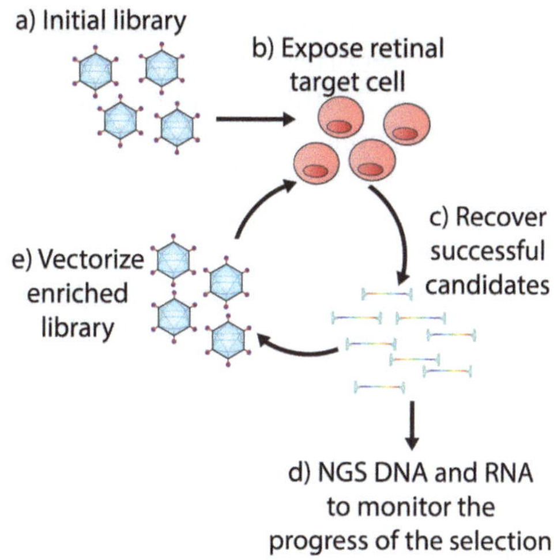

Fig. 5 Overview of capsid library selection. (**a**) The initial library with high variability is used to transduce the target cells. Step **a** is only performed once, while all other steps are performed as iterative cycles. (**b**) Target cells are exposed with the library and cultured depending on their requirements. (**c**) Successful capsids are recovered, usually in the form of viral vector DNA, and other methods are possible. (**d**) Progress of the selection is monitored by deep sequencing (NGS). (**e**) An enriched capsid library is generated from the recovered genetic information of successful candidates. The first round of selection comprises steps **a**–**d**, while every subsequent round of selection is only comprised of steps **b**–**e**

Both DNA and RNA/cDNA are amplified, and the VR-VIII region containing the library is analyzed by Illumina deep sequencing (Fig. 5d). This enriched library can be re-vectorized (Fig. 5e), and the target cells will be transduced with the enriched library.

For the purpose of this protocol, the exact model in which the capsid library directed evolution is performed has no effect on the selection. The library can be used for:

1. In vivo selection in rodents (intravitreal or subretinal injection) [5, 6]
2. In vivo selection in nonhuman primate (intravitreal, subretinal, or suprachoroidal injection) [4]
3. Selection in primary human retina explants in interphase or immersion culture [16]
4. Selection in stem-cell-derived retina organoids [17]

The culture/animal handling conditions should be chosen to keep viability optimal, and the dose of AAV library should be comparable to previously used doses and durations of AAV exposure. A good duration to start with is the time it takes for present benchmark for the given cell type to show visible fluorescent reporter transgene expression.

After the appropriate duration of AAV exposure, relevant cells should be harvested using the most appropriate methods, and DNA and RNA need to be extracted and processed for deep sequencing and enriched library cloning.

Detailed Steps for the Library Recovery and Subsequent Cloning

See **Note 10**.

1. Perform DNA and RNA extraction using the AllPrep DNA/RNA Kit following the manufacturer's instructions (*see* **Note 11**).
2. DNA samples can be stored at −20 °C. **Steps 78–87** are for RNA only.
3. Perform residual DNA digest using TURBO DNase in a 0.5 mL tube:
   (a) 500 ng RNA.
   (b) 2 μL TURBO DNase buffer.
   (c) 1 μL TURBO DNAse.
   (d) Fill to 20 μL with nuclease-free water.
4. Incubate at 37 °C for 60 min. Add further 1 μL of TURBO DNase, mix, and incubate another 60 min.
5. Add 4 μL inactivation reagent, and incubate for 5 min at room temperature.
6. Spin at 10,000 × g for 5 min at room temp and carefully transfer 20 μL into PCR tube.
7. To 20 μL DNase-treated RNA add:
   (a) 3 μL oligo-dT primer
   (b) 3 μL dNTPs
   (c) 4 μL nuclease-free water
8. Heat RNA–primer mix at 80 °C for 1 min and 65 °C for 5 min. Then incubate on ice for 1–5 min.
9. Divide into three 10 μL samples:
   (a) 10 μL RT+
   (b) 10 μL RT-
   (c) 10 μL backup
10. To RT+ sample add:
    (a) 4 μL SSIV buffer
    (b) 1 μL 100 mM DTT
    (c) 1 μL RNAseOUT Recombinant RNAse Inhibitor

(d) 1 μL SuperScript IV reverse transcriptase
(e) 3 μL nuclease-free water

11. To RT- sample add:
    (a) 4 μL SSIV buffer
    (b) 1 μL 100 mM DTT
    (c) 1 μL RNAseOUT Recombinant RNAse Inhibitor
    (d) 4 μL nuclease-free water
12. Incubate reaction at 53 °C for 30 min, and inactivate at 80 °C for 10 min.
13. The following steps apply to DNA from **step** 77 and cDNA from **step 87**. Perform PCR amplification of the peptide-encoding region (*see* **Note 11**). For each reaction, in a total volume of 50 μL mix:
    (a) 5 μL cDNA (RT+ and RT-) or DNA sample (20 ng/μL) or vector input library
    (b) 2.5 μL 10 mM peptide forward primer ACTAACCCTGTGGCCACGG
    (c) 2.5 μL 10 mM peptide reverse primer GGACCCTGCAGGTACACGTC
    (d) 10 μL Q5 reaction buffer
    (e) 1 μL dNTPs
    (f) 0.5 Q5 HotStart DNA polymerase
    (g) 28.5 μL nuclease-free water
14. Run the PCR reaction using the following cycles:
    (a) 98 °C for 10 min
    (b) 30–42 cycles:
        (i) 98 °C for 10 s
        (ii) 69 °C for 15 s
        (iii) 72 °C for 5 s
    (c) 72 °C for 2 min
15. Run PCR on 2% agarose gel. Expected product size: 167 bp.
16. Extract bands from agarose gel, purify, and submit for PE150 Illumina deep sequencing. Always retain 250 ng of amplicon for enriched library cloning (*see* **Note 12**).
17. The library cloning is performed in a similar way as the original library (**steps 24–37**), only that the library insert is now the PCR product from the round 1 selection.
18. Set up a large-scale assembly reaction separately for the secondary/enriched libraries by mixing:
    (a) 987 ng (225 fmol) Sfi-digested pFT-Cap2-SfiI (same as before)

(b) 232 ng (2250 fmol) library amplicon from **RT+ cDNA** (more as the amplicon is longer than the dsDNA oligo) (*see* **Note 14**).

(c) Fill up to 80 μL with nuclease-free water.

(d) Add 80 μL NEBuilder HiFi DNA assembly and mix well.

(e) Separate into eight PCR tubes and incubate at 50 °C for 30 min.

19. Desalting: Combine all eight reactions into one 1.5 mL tube and add 16 μL 3 M sodium acetate and 400 μL 100% ethanol, analytical grade. Incubate mix at −80 °C for 30 min (*see* **Note 4**).
20. Perform electroporation as described above (**steps 32–37**).

Following the outlined steps, the library can be selected for several more rounds. The point when to stop with the iterative rounds of selection needs to be decided based on the deep sequencing results. Once a suitable level of enrichment is achieved, the most successful variants need to be cloned into the pRep2Cap2 plasmid used at the very first step of this protocol. Depending on the library enrichment results, a few selected variants can be vectorized and tested individually, using relevant benchmark AAV as references. The benchmarks can be selected from the literature, or using AAV Testing Kit [18]. Alternatively, a larger number of successful AAV capsids can be selected for vectorization and compared to one another and benchmarks using a multiplexed barcoded approach previously published [18].

4 Notes

1. The HF version of NsiI is not compatible with the buffer 3.1, and using it will make the process more complicated.
2. Complete digest and dephosphorylation of the pFT-Cap-SfiI plasmid with SfiI is important for the correct cloning of the library. If the digest is incomplete, and even a very small amount of undigested plasmid is present, the carryover of religated plasmid without library will be very high as the NEBuilder mix has a ligase activity. Based on our results from deep sequencing analysis, long digests and dephosphorylation are critical and effective to completely ablate any religation.
3. Oligonucleotides purified with a more stringent method, such as HPLC or PAGE, will reduce the oligonucleotide pool diversity.
4. Desalting by ethanol precipitation of the assembly reaction is extremely important to ensure successful electroporation, as the osmolarity of the bacterial DNA solution that undergoes

electroporation must be sufficient to transmit the pulse between the electrodes, but not too high as that can kill the bacteria.

5. In the authors' experience, a diversity of 10^7 colonies after transformation yielded a complexity of up to 5×10^6 capsid variants based on NGS results of the preselection library, which we consider a highly diverse functional library.
6. We are using trimethoprim to perform liquid selection of successfully transformed bacterial clones. This approach is more efficient than performing a plating selection with ampicillin-resistant clones on hundreds of LB-agar plates.
7. All amounts are calculated with 10% contingency. Perform all steps in a biosafety cabinet to ensure sterility.
8. If none of the tubes used in the same spin have a visible band, use a Parafilm to seal the top of the tube, poke a hole at the bottom of the tube, and collect 0.5 mL fractions. Gently invert tubes ~5 times to mix for fraction analysis. To perform analysis of individual fractions: (A) Take 1 μL from each sample and dilute in 5 μL dH2O (in 200 μL PCR tube) for SDS silver stain. (B) Take 1 μL from each fraction and dilute in 19 μL dH2O (in 200 μL PCR tube) for dPCR/ddPCR/qPCR analysis. (C) Determine which fraction has the highest AAV-VP1/2/3 protein content and genomic titer to decide which fractions (up to 3 mL) should be pooled.
9. Clean every surface, items, and devices carefully with RNase away prior to use. All RNA work should be performed on ice unless stated otherwise.
10. Several different ways to extract and process DNA and RNA are available. We prefer to use the DNA/RNA AllPrep Kit as it enables the DNA and RNA extraction from the same sample.
11. For RT+ perform 2× PCR reactions. For RT- control, only one. Always include NTC reaction. The peptide primers work for all local codon optimized serotypes apart from AAV4, AAV5, AAV11, and AAV12.
12. As long as the sequencing is based on PE150 Illumina amplicon-seq, the results should not differ depending on the provider of the sequencing services.

Author Contributions A.W. and L.L. wrote the manuscript and generated the figures. All authors reviewed, edited, and commented on the manuscript.

Author Disclosure Statement L.L. has commercial affiliations. L.L. has consulted on technologies discussed in this protocol. L.L. has stock and/or equity in companies with technologies broadly related to this topic. L.L. and A.W. are co-inventors of AAV technologies described in this protocol. All other authors declare no competing financial interests.

Conflict of Interest The authors have no conflicts of interest to declare that are relevant to the content of this chapter.

References

1. Buck TM, Wijnholds J (2020) Recombinant adeno-associated viral vectors (rAAV)-vector elements in ocular gene therapy clinical trials and transgene expression and bioactivity assays. Int J Mol Sci 21(12). https://doi.org/10.3390/ijms21124197
2. Bennett J (2017) Taking stock of retinal gene therapy: looking Bback and moving forward. Mol Ther 25(5):1076–1094. https://doi.org/10.1016/j.ymthe.2017.03.008
3. Russell S, Bennett J, Wellman JA, Chung DC, Yu Z-F, Tillman A, Wittes J, Pappas J, Elci O, McCague S, Cross D, Marshall KA, Walshire J, Kehoe TL, Reichert H, Davis M, Raffini L, George LA, Hudson FP, Dingfield L, Zhu X, Haller JA, Sohn EH, Mahajan VB, Pfeifer W, Weckmann M, Johnson C, Gewaily D, Drack A, Stone E, Wachtel K, Simonelli F, Leroy BP, Wright JF, High KA, Maguire AM (2017) Efficacy and safety of voretigene neparvovec (AAV2-hRPE65v2) in patients with RPE65-mediated inherited retinal dystrophy: a randomised, controlled, open-label, phase 3 trial. Lancet 390(10097):849–860. https://doi.org/10.1016/S0140-6736(17)31868-8
4. Byrne LC, Day TP, Visel M, Strazzeri JA, Fortuny C, Dalkara D, Merigan WH, Schaffer DV, Flannery JG (2020) In vivo–directed evolution of adeno-associated virus in the primate retina. JCI Insight 5(10). https://doi.org/10.1172/jci.insight.135112
5. Dalkara D, Byrne LC, Klimczak RR, Visel M, Yin L, Merigan WH, Flannery JG, Schaffer DV (2013) In vivo–directed evolution of a new adeno-associated virus for therapeutic outer retinal gene delivery from the vitreous. Sci Transl Med 5(189):189ra176. https://doi.org/10.1126/scitranslmed.3005708
6. Klimczak RR, Koerber JT, Dalkara D, Flannery JG, Schaffer DV (2009) A novel adeno-associated viral variant for efficient and selective intravitreal transduction of rat Muller cells. PLoS One 4 (10):e7467. https://doi.org/10.1371/journal.pone.0007467
7. Westhaus A, Cabanes Creus M, Jonker T, Sallard E, Navarro RG, Zhu E, Baltazar G, Lee S, Wilmott P, Gonzalez-Cordero A, Santilli G, Thrasher AJ, Alexander IE, Lisowski L (2022) AAV-p40 bioengineering platform for variant selection based on transgene expression. Hum Gene Ther 33(11–12):664–682. https://doi.org/10.1089/hum.2021.278
8. Müller OJ, Kaul F, Weitzman MD, Pasqualini R, Arap W, Kleinschmidt JA, Trepel M (2003) Random peptide libraries displayed on adeno-associated virus to select for targeted gene therapy vectors. Nat Biotechnol 21(9): 1040–1046. https://doi.org/10.1038/nbt856
9. Börner K, Kienle E, Huang L-Y, Weinmann J, Sacher A, Bayer P, Stüllein C, Fakhiri J, Zimmermann L, Westhaus A, Beneke J, Beil N, Wiedtke E, Schmelas C, Miltner D, Rau A, Erfle H, Kräusslich H-G, Müller M, Agbandje-McKenna M, Grimm D (2020) Pre-arrayed pan-AAV peptide display libraries for rapid single-round screening. Mol Ther. https://doi.org/10.1016/j.ymthe.2020.02.009
10. Cabanes-Creus M, Ginn SL, Amaya AK, Liao SHY, Westhaus A, Hallwirth CV, Wilmott P, Ward J, Dilworth KL, Santilli G, Rybicki A, Nakai H, Thrasher AJ, Filip AC, Alexander IE, Lisowski L (2019) Codon-optimization of wild-type adeno-associated virus capsid sequences enhances DNA family shuffling while conserving functionality. Mol Ther 12: 71–84. https://doi.org/10.1016/j.omtm.2018.10.016
11. Deverman BE, Pravdo PL, Simpson BP, Kumar SR, Chan KY, Banerjee A, Wu W-L, Yang B, Huber N, Pasca SP, Gradinaru V (2016) Cre-dependent selection yields AAV variants for widespread gene transfer to the adult brain. Nat Biotechnol 34:204. https://doi.org/10.1038/nbt.3440. https://www.nature.com/articles/nbt.3440#supplementary-information

12. Hanlon KS, Meltzer JC, Buzhdygan T, Cheng MJ, Sena-Esteves M, Bennett RE, Sullivan TP, Razmpour R, Gong Y, Ng C, Nammour J, Maiz D, Dujardin S, Ramirez SH, Hudry E, Maguire CA (2019) Selection of an efficient AAV vector for robust CNS transgene expression. Mol Ther 15:320–332. https://doi.org/10.1016/j.omtm.2019.10.007
13. Nonnenmacher M, Wang W, Child MA, Ren X-Q, Huang C, Ren AZ, Tocci J, Chen Q, Bittner K, Tyson K, Pande N, Chung CH-Y, Paul SM, Hou J (2021) Rapid evolution of blood-brain-barrier-penetrating AAV capsids by RNA-driven biopanning. Mol Ther 20: 366–378. https://doi.org/10.1016/j.omtm.2020.12.006
14. Nakai H, Huang S, Adachi K (2020) Methods for using transcription-dependent directed evolution of aav capsids. Patent US20230119163A1. Published 2023-04-20
15. Schmit PF, Pacouret S, Zinn E, Telford E, Nicolaou F, Broucque F, Andres-Mateos E, Xiao R, Penaud-Budloo M, Bouzelha M, Jaulin N, Adjali O, Ayuso E, Vandenberghe LH (2020) Cross-packaging and capsid mosaic formation in multiplexed AAV libraries. Mol Ther 17:107–121. https://doi.org/10.1016/j.omtm.2019.11.014
16. Westhaus A, Eamegdool SS, Fernando M, Fuller-Carter P, Brunet AA, Rashwan R, Knight M, Daniszewski M, Lidgerwood GE, Pébay A, Hewitt A, Santilli G, Thrasher AJ, Carvalho LS, Gonzalez-Cordero A, Jamieson RV, Lisowski L (2023) AAV capsid bioengineering in primary human retina. Research Square (Preprint). https://doi.org/10.21203/rs.3.rs-3293367/v1
17. Gonzalez-Cordero A, Goh D, Kruczek K, Naeem A, Fernando M, kleine Holthaus S-M, Takaaki M, Blackford SJI, Kloc M, Agundez L, Sampson RD, Borooah S, Ovando-Roche P, Mehat MS, West EL, Smith AJ, Pearson RA, Ali RR (2018) Assessment of AAV vector tropisms for mouse and human pluripotent stem cell–derived RPE and photoreceptor cells. Hum Gene Ther 29(10):1124–1139. https://doi.org/10.1089/hum.2018.027
18. Westhaus A, Cabanes-Creus M, Rybicki A, Baltazar G, Navarro RG, Zhu E, Drouyer M, Knight M, Albu RF, Ng BH, Kalajdzic P, Kwiatek M, Hsu K, Santilli G, Gold W, Kramer B, Gonzalez-Cordero A, Thrasher AJ, Alexander IE, Lisowski L (2020) High-throughput in vitro, ex vivo, and in vivo screen of adeno-associated virus vectors based on physical and functional transduction. Human Gene Therapy Methods. https://doi.org/10.1089/hum.2019.264

Chapter 2

Evaluation of CRISPR/CasRx-Mediated VEGF mRNA Knockdown in Mouse Retina

Satheesh Kumar, Da Zhao, Vickie H. Y. Wong, Bang V. Bui, and Guei-Sheung Liu

Abstract

Neovascular eye diseases (NEDs) are a group of diseases caused by the abnormal overgrowth of blood vessels in the eye. Normal vasculature is maintained through a dynamic balance of the vascular endothelial growth factor (VEGF). In diseases such as advanced diabetic retinopathy (DR) and neovascular age-related macular degeneration (AMD), an overexpression of VEGF leads to the formation of structurally weak and leaky blood vessels, resulting in vision impairment and, without intervention, legal blindness. In the clinic, NEDs are presently managed through anti-VEGF agents that specifically bind and neutralize VEGF signaling. While effective, this approach requires invasive intravitreal injections monthly and places a heavy burden on patients and healthcare providers. A flexible, long-lasting therapeutic that can reduce or eliminate frequent anti-VEGF treatment will significantly advance NED management.

RNA editing with clustered regularly interspaced short palindromic repeats (CRISPR)-associated protein (Cas) is an emerging strategy to achieve reversible gene editing. The CRISPR-Cas13 system exclusively targets single-stranded RNA and allows gene silencing in a safe manner through RNA knockdown, as the genome is left intact. In addition, the compact CasRx enzyme (930aa) allows RNA silencing to be achieved through the delivery of a single adeno-associated virus (AAV), ideal for gene therapy applications. Herein, we outline methods to target *VEGFA* mRNA using CRISPR/CasRx in a mammalian cell line and humanized VEGFA transgenic mice.

Key words CRISPR-Cas, RNA editing, Retinal neovascularization, Anti-VEGF, Gene therapy, AAV, Age-related macular degeneration, Diabetic retinopathy

1 Introduction

Neovascular eye diseases (NEDs), such as advanced diabetic retinopathy (DR) and neovascular age-related macular degeneration (AMD), are sight-threatening conditions that remain a leading cause of vision impairment and blindness globally. They progress through neuronal and vascular dysfunction, followed by neovascularization which impairs vision [1]. Neovascularization is heavily influenced by the vascular endothelial growth factor (VEGF) and is

Livia S. Carvalho (ed.), *Retinal Gene Therapy: Methods and Protocols*, Methods in Molecular Biology, vol. 3023,
https://doi.org/10.1007/978-1-0716-5198-8_2,
© The Author(s), under exclusive license to Springer Science+Business Media, LLC, part of Springer Nature 2026

thus a common target in therapeutic development [2]. Presently, NEDs are managed through the administration of anti-VEGF drugs, which use small molecules or monoclonal antibodies to target and neutralize VEGF. As NEDs are chronic conditions, there is a need for frequent administration of anti-VEGF drugs, and this is often achieved through monthly or bimonthly intravitreal injections in the clinic. In addition to carrying a heavy financial burden, repeated ocular injections are greatly distressful to patients and laborsome to clinicians. Long-term therapeutic options are therefore being explored for alternative intervention strategies.

The advent of gene technology has recently allowed researchers to explore gene-targeted therapeutics (gene therapy). Notably, the clustered regularly interspaced short palindromic repeats (CRISPR) system has revolutionized gene therapy through achieving precise genomic modifications [3]. Coupled with CRISPR-associated (Cas) enzymes, CRISPR-Cas can achieve targeted insertions and deletions in the genome. With recent developments in CRISPR/Cas technology, transcriptome engineering has also been demonstrated, offering an even safer alternative to target genes without inducing permanent changes in the genome [4]. Importantly, unlike most DNA-targeting CRISPR/Cas systems, several RNA-targeting CRISPR systems allow transcriptome-wide targeting and are adequately compact for adeno-associated virus (AAV)-mediated delivery [5], making them ideal for therapeutic development.

In this chapter, we outline methods for designing and demonstrating highly efficient VEGFA mRNA knockdown using the CRISPR/CasRx system in a mammalian cell line. In addition, we describe intravitreal delivery of all-in-one AAV.CRISPR/CasRx into Kimba transgenic mice and analysis of retinal vasculature and pathology through fundus fluorescence angiography (FFA) and optical coherence tomography (OCT). Overall, this provides a comprehensive manual for demonstrating AAV-compatible RNA silencing using CRISPR/CasRx and detailed methods to analyze retinal vascular pathology.

2 Materials

2.1 sgRNA Design and Vector Construction

1. Cas13 design: https://cas13design.nygenome.org/.
2. CRISPR-RfxCas13d (CasRx) sequence was a gift from Dr. Patrick Hsu (Salk Institute for Biological Studies) [5].
3. Addgene plasmids: pAAV.CMV-mCherry (Addgene ID: 205744); pAAV.CMV-CasRx-T2A-mCherry (Addgene ID: 203442); pAAV.CMV-CasRx-VEGFA sgRNA (Addgene ID: 203443); pAAV.CMV-CasRx-VEGFA sgRNA array (Addgene ID: 203444).
4. Vector construction using Benchling: https://benchling.com/.

2.2 Cell Culture

1. Human Embryonic Kidney (HEK) 293FT cell line.
2. Cell culture media (complete DMEM): Dulbecco's modified Eagle's medium (DMEM) high glucose, 10% v/v fetal bovine serum, 1% v/v antibiotic–antimycotic, 1% v/v GlutaMAX Supplement.
3. Falcon® 75 cm^2 rectangular canted neck cell culture flask with vented cap.
4. Multiwell cell culture plates.
5. Dulbecco's phosphate-buffered saline (DPBS), no calcium, no magnesium.
6. TryPLE™ Express Enzyme (1×), phenol red.

2.3 Transfection and RNA Extraction

1. Opti-MEM™ Reduced Serum Medium.
2. Lipofectamine™ 3000 Transfection Reagent or equivalent.
3. RNA extraction kit.

2.4 Reverse Transcription and Quantitative Polymerase Chain Reaction

1. High-Capacity cDNA Reverse Transcription Kit.
2. TaqMan™ Fast Advanced Master Mix or equivalent.
3. Human VEGFA TaqMan™ Fluorescent Real-Time PCR Primers (Hs00900055_m1).
4. Human GAPDH TaqMan™ Fluorescent Real-Time PCR Primers (Hs99999905_m1).

2.5 Equipment

1. Tissue culture equipment (e.g., cell incubators, cell counter, water baths, and microscopes).
2. Centrifuges (e.g., high-speed centrifuge and microcentrifuge).
3. NanoDrop One.
4. PCR thermal cycler.
5. Quantitative polymerase chain reaction (qPCR) instrument.

2.6 RNA Sequencing Analysis

1. RStudio (https://posit.co/download/rstudio-desktop/).
2. Galaxy (https://usegalaxy.org/).
3. Bioconductor package: DESeq2.
4. GENCODE Annotation files (https://www.gencodegenes.org/human/).

2.7 Intravitreal Injection

1. Kimba ($VEGFA^{+/+}$) mice (trVEGF029; licensed from Lions Eye Institute Ltd., Prof Elizabeth Rakoczy) [6].
2. 10 μL Hamilton syringe or equivalent.
3. 33-gauge dental needles.
4. 0.28 mm polyethylene tubing.

5. Heating pads.
6. Ketamine–xylazine.
7. Atipamezole hydrochloride (5 mg/mL).

2.8 Fluorescent Fundus Angiography (FFA) and Optical Coherence Tomography (OCT)

1. Topical proxymetacaine (0.5% Alcaine): topical anesthetic.
2. 1% tropicamide: pupil dilation.
3. Sodium fluorescein.
4. AngioTool software (https://ccrod.cancer.gov/confluence/display/ROB2/Home).
5. HEYEX2 software.
6. Ocular optical coherence tomography with adapted for rodent imaging such as the Spectralis OCT2 or equivalent.
7. 3 mg hypromellose and 2 mg carbomer 980 lubricating eye drops and eye gel.

3 Methods

3.1 Design of sgRNA and Vector Construction

1. Go to: https://cas13design.nygenome.org/.
2. Search for *VEGFA* gene under "Lookup human transcript expression."
3. Retrieve transcript ID, select "Human" tab, and insert into the search bar.
4. Identify three top scoring gRNAs in the coding region ("CDS" under Annotation).
5. Plasmids can be designed using Benchling (https://www.benchling.com/).
6. Clone gRNAs into vector backbone using Gibson assembly as previously described [7].
7. For *VEGFA* targeting constructs, obtain Addgene plasmids for single gRNA and gRNA array constructs, respectively (Fig. 1a).

3.2 Transfection and RNA Extraction of HEK293FT Cells with All-in-One pAAV. CRISPR/CasRx Constructs

1. The day before transfection, seed 3×10^5 HEK293FT cells per well of 12-well plates in complete DMEM culture media incubated at 37 °C with 5% CO_2 (*see* **Note 1**). The following day, change media before transfection procedure.
2. Set up transfection reaction with 1500 ng of each plasmid and 3 μL of P3000 reaction in 50 μL Opti-MEM for each well.
3. Add 4.5 μL of Lipofectamine 3000 (or equivalent transfection reagent) in 50 μL of Opti-MEM for each well.
4. Mix plasmid and Lipofectamine well and incubate in room temperature for 15 min for DNA transfection complex formation.

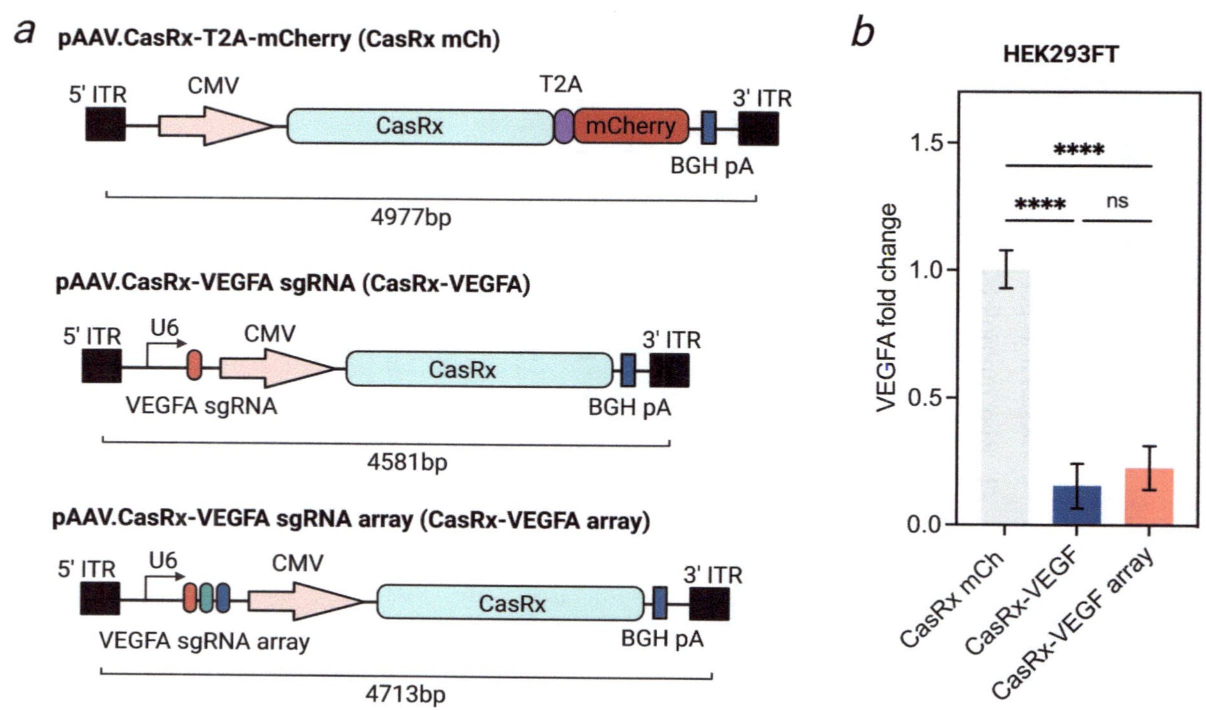

Fig. 1 (**a**) Domain structure of all-in-one pAAV.CasRx plasmids. (**b**) Level of VEGFA mRNA knockdown from all-in-one pAAV.CasRx plasmids

5. Add 100 μL of the transfection mix dropwise into each well, swirl to mix into medium, and incubate plate in 37 °C with 5% CO_2.
6. Two days after transfection, image for fluorescence to determine transfection efficiency.
7. Harvest the cells for RNA extraction using an RNA extraction kit following the manufacturer's protocol and eluting in 50 μL of RNase-free water (*see* **Note 2**).

3.3 Multiplex qPCR and RNA Knockdown

1. Perform reverse transcription using High-Capacity cDNA synthesis kit as per manufacturer's instructions (*see* **Note 3**).
2. Run multiplex qPCR reaction with TaqMan human VEGFA and GAPDH primers as previously described [7].
3. Determine RNA knockdown rate using $2^{-\Delta\Delta Ct}$ and normalize to control (Fig. 1c).
4. If knockdown efficiency is unsatisfactory, check transfection efficiency or plasmid integrity (*see* **Note 4**).

3.4 RNA Sequencing and Analysis

1. Perform RNA sequencing on harvested RNA from preferred sequencing facility (e.g., Azenta Life Sciences). Read sequences are obtained as FASTQ files.
2. Go to https://usegalaxy.org/ for RNA-Seq pipeline (Fig. 2).
3. Go to "Upload Data" and upload FASTQ files.

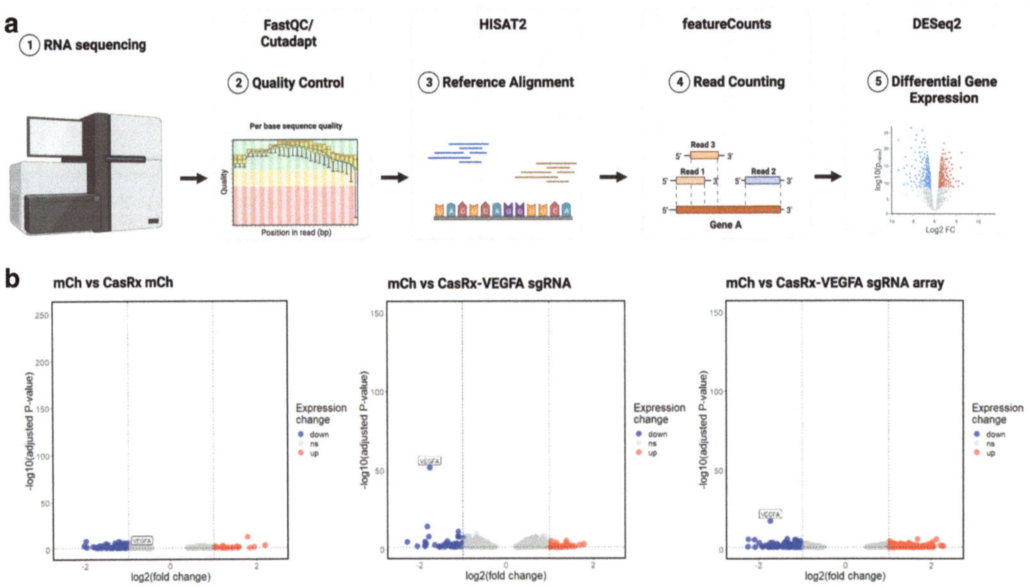

Fig. 2 (**a**) Schematic of RNA sequencing analysis pipeline. (**b**) Volcano plots for differentially expressed genes of CasRx mCh (pAAV.CasRx-T2A-mCherry), CasRx VEGF (pAAV.CasRx-VEGFA sgRNA), and CasRx VEGF array (pAAV.CasRx-VEGFA sgRNA array) compared to transfection control (pAAV.mCherry)

4. Search for "FastQC' package and perform initial quality control.
5. Open "Cutadapt" for trimming adaptors and removing low-quality reads.
6. Input uploaded FASTQ files and execute.
   (a) In *"Filter Options"*: *"Minimum length (R1)"*: 20.
   (b) In *"Read Modification Options"*: *"Quality cutoff"*: 20.
7. Open "HISAT2" for sequence alignment to human genome (output will be a SAM/BAM file).
   (a) Input trimmed FASTQ file.
   (b) Select "Human (*Homo sapiens*) (b38): hg38" for reference genome and execute.
8. Go to https://www.gencodegenes.org/human/ and download GTF file for "Basic Gene Annotation" for all regions and upload to Galaxy.
9. Open "featureCounts," and input aligned SAM/BAM file.
   (a) Under "Gene Annotation File," select "A GFF/GTF file in your history."
   (b) Choose uploaded GTF file.
   (c) Run tool.
   (d) Download output counts file as "txt" file.
10. Open RStudio.
11. Access counts file in R:

```
countdata <- read.table("counts_gencode_batch1.txt",header=-
TRUE,row.names=1)
```

12. Trim the data for differential gene expression analysis:

```
# Remove first five columns (chr,start,end,strand,length)
countdata <- countdata[ ,6:ncol(countdata)]
# Remove.bam or.sam from filenames
colnames(countdatai <- gsub("//.[sb]am$","",colnames(-
countdata))
# Convert to matrix
countdatal <- as.matrix(countdata
```

13. Select control and treatment conditions and number of biological repeats:

```
# Assign conditions
(condition <- factor(c(rep("CasRx NT",3),rep("CasRx
VEGF",3))))
```

14. Install DESeq2 package:

```
BiocManager::install("DESeq2")
library(DESeq2)
```

15. Run DESeq2:

```
# Create a coldata frame and instantiate the DESeqDataSet.
(coldata <- data.frame(row.names=colnames(countdata1),condi-
tion))
dds <- DESeqDataSetFromMatrix(countData=countdata1,colData
=coldata,design=~condition)
# Run the DESeq pipeline
res <- DESeq(dds)
```

16. Plot differentially expressed genes as a volcano plot (Fig. 2b–d):

```
# Volcano Plot
volcanoplot <- function (res,lfcthresh=2,sigthresh=0.05,xlab
="log2(Fold Change)",legendpos="topright",labelsig
=FALSE,textcx=1.5,...) {
with(res,plot(log2FoldChange,-log10(pvalue),pch=20,xlab=x-
lab,cex.axis
=1.8,cex.lab=1.5,...))
with(subset(res,padj<sigthresh),points(log2FoldChange,-log10
(pvalue),pch
=20,col="blue",...))
with(subset(res,abs(log2FoldChange)
>lfcthresh),points(log2FoldChange,-log10(pvalue),pch=20,
col="orange",...))
with(subset(res,padj<sigthresh & abs(log2FoldChange)
>lfcthresh),points(log2FoldChange,-log10(pvalue),pch=20,col
="green",...))
legend(legendpos,xjust=1,yjust=1,legend=c(paste("p-adj
<",sigthresh,sep=""),paste("|log2(FC)|>",lfcthresh,sep
=""),"both"),cex=1.5,pch=20,col=c("blue","orange","green"))
}
volcanoplot(resdata,lfcthresh=2,sigthresh=0.05,xlim=c(-6,6),
ylim
=c(0,33),legendpos="topright"
```

3.5 Identification of Top Off-Target Genes and Guide-Dependent Off-Target Activity

1. Go to NCBI BLAST (https://blast.ncbi.nlm.nih.gov/Blast.cgi).
2. Select "Nucleotide BLAST."
3. Input sgRNA sequence:

   (a) Under "Database": Choose "Standard Databases"; "RefSeq Standard RNA sequences."
   (b) Under "Program Selection": Choose "Somewhat similar sequences."
   (c) Under "General Parameters": Choose Word Size = 15.
   (d) Run BLAST.
4. Click on "Query Cover" to sort results by sequence alignment.
5. Record top 10 genes after target gene.
6. Obtain read counts from RNA-Seq analysis for the respective genes.
7. Calculate transcripts per million (TPM) to determine abundance of gene:
   (a) Divide the read counts by gene length in kilobases. This gives you reads per kilobase (RPK).
   (b) Sum all reads in a sample and divide this number by 1,000,000. This is your "per million" scaling factor.
   (c) Divide the RPK values by the "per million" scaling factor to obtain TPM.
8. Compare TPM values for control and treated samples to determine guide dependent off-target activity.

3.6 In Vivo CRISPR/CasRx VEGFA Knockdown in Kimba Mice

1. Mouse model of retinal neovascularization: Kimba mice (trVEGF029) are humanized transgenic mice that overexpress human *VEGFA* from photoreceptor cells [6]. The animal shows early upregulation of VEGFA protein coinciding with early damage to vasculature and neural cells. Importantly, the model mimics human DR through neovascularization mediated by overexpression of VEGF and structural damage to the choroid. Vascular lesions also first appear centrally, close to the optic nerve, as in most mammals along with microaneurysms identical to human DR.
2. As Kimba mice offer a promising model for investigating therapies for DR, we sought to target human VEGF in Kimba photoreceptors to demonstrate potential for in vivo CasRx knockdown. To this end, we packaged our pAAV.CasRx plasmids in AAV2.7m8 capsid to investigate *VEGFA* mRNA knockdown in vivo. In addition, AAV2.7m8 at high doses can achieve good transduction of mice and nonhuman primate photoreceptors from intravitreal delivery, a minimally invasive and commonly performed procedure in humans [8].

3.7 Intravitreal Injection

1. Weigh mouse to determine the volume of anesthetic.
2. Anesthetize mouse using IP injection of ketamine–xylazine (80:10 mg/kg) and allow mouse to immobilize (takes about 5 minutes).

3. Keep the animal warm throughout the injection using the heat pad.
4. Once immobilized, place mouse under a dissection microscope for injection.
5. Connect a 33-gauge dental needle onto a 10 μL Hamilton syringe or equivalent using 0.28 mm inner diameter PE tubing.
6. Prime the line to remove air using sterile saline.
7. Draw 0.5 μL of air followed by 1 μL of virus suspension for injection.
8. Puncture the sclera in the superior temporal quadrant of the eye at 1 mm behind the limbus using a 33-gauge needle.
9. Insert the 33-gauge needle into vitreous cavity (~1 mm) and gently inject virus suspension and air into the eye (*see* **Note 5**).
10. Maintain the needle inside the eye for further 20 s to minimize fluid reflux.
11. Remove needle from eye and observe for 10 s for bleeding.
12. Perform subcutaneous injection of 0.2–0.4 mL atipamezole hydrochloride to reverse the effect of ketamine–xylazine anesthesia, and place mouse on heating pad for recovery.

3.8 Fluorescence Fundus Angiography (FFA) and Optical Coherence Tomography (OCT) Imaging

1. Anesthetize mouse as previously described (*see* Subheading 3.6, **steps 1** and **2**).
2. Place one drop of topical proxymetacaine and tropicamide to each eye and place eye gel with cover slips to prevent drying.
3. Perform IP injection of fluorescein (4 mg/kg) and immediately position the animal on the stage with a heat pad (to maintain body temperature) in front of the Spectralis or equivalent for FFA/OCT imaging.
4. Hydrate corneas using lubricating eye drops prior to and constantly during imaging.
5. For FFA, acquire central image over the optic nerve head and of the four quadrants (*see* **Note 6**).
6. For OCT, acquire images with a volumetric scan pattern (7.6 × 6.3 × 1.9 mm) centered over the optic nerve head. Each volume scan will consist of 121 vertical B-scans (five repeats), and each B-scan made up of 768 A-scans.

3.9 Analysis of Retinal Vasculature from Fluorescence Fundus Angiography (FFA)

1. Open AngioTool software [9].
2. Click on "Open Image" and open first FFA image to be analyzed (Fig. 3, step 1).
3. Click on "Run Analysis." AngioTool automatically identifies vascular network and quantifies several parameters (e.g., vessel density, vessel length, lacunarity, etc.) (Fig. 3, step 2).

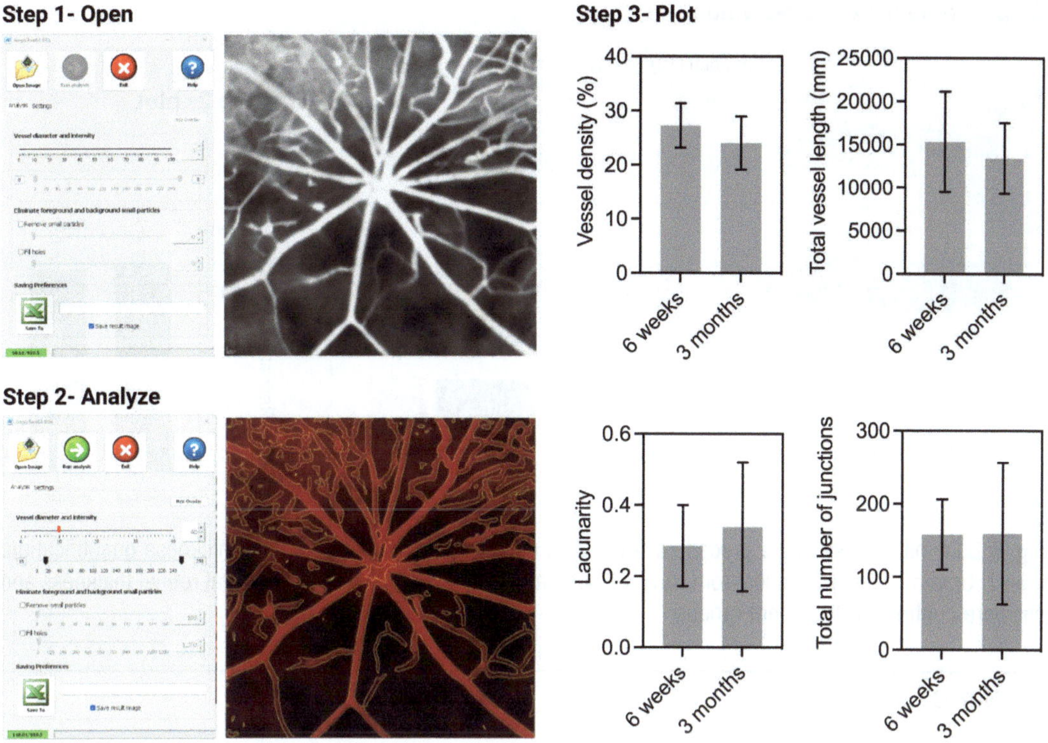

Fig. 3 Schematic of stepwise procedure for AngioTool analysis of mouse retinal vasculature. Step 1: Open image to be analyzed. Step 2: Click on Run analysis. Step 3: Collect relevant parameters (vessel density, lacunarity, etc.) and plot and compare between groups

4. Repeat analysis for subsequent images.
5. Average values for vessel density and lacunarity for each eye.
6. Plot averaged values to determine vascular growth (Fig. 3, step 3).

3.10 Analysis of Retinal Thickness from OCT

1. Open HEYEX2 software.
2. Browse for first OCT image and go to "Thickness Map" (Fig. 4, step 1).
3. The Early Treatment of Diabetic Retinopathy Study (EDTRS) grid can be used as a guide to obtain averaged thickness values from nasal and temporal sides of the retina.
4. Collect N2 (outer segment of nasal) and T2 (outer segment of temporal) values.
5. Repeat **step 1** for each eye.
6. Plot nasal and temporal thickness for all animals to determine change in retinal thickness (Fig. 4, step 2).

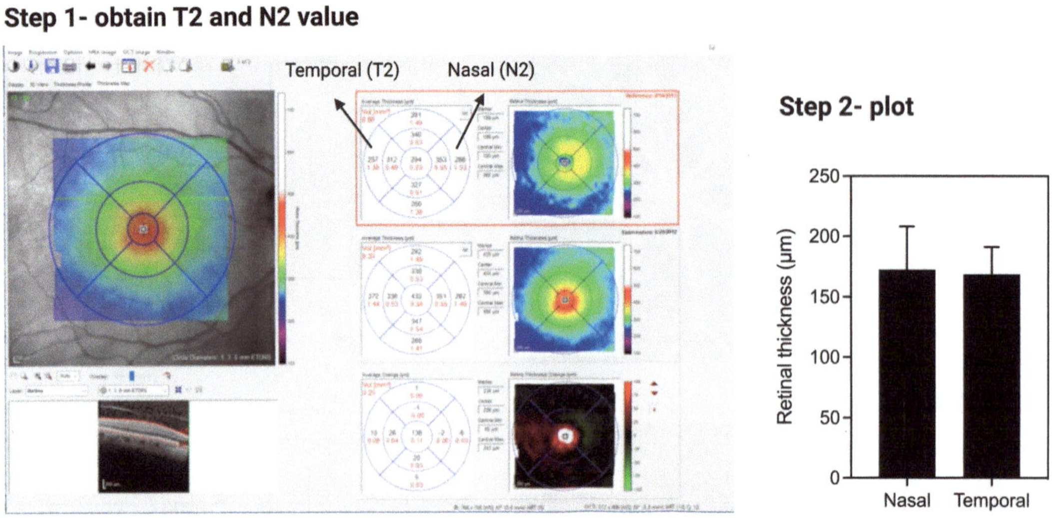

Fig. 4 Schematic of stepwise procedure for optical coherence tomography retinal thickness quantification. Step 1: Obtain T2 and N2 values from HEYEX 2 software. Step 2: Plot nasal and temporal retinal thickness and compare control and treatment groups

4 Notes

1. For best transfection results, do not allow cells to become too confluent before using the cells for transfection. Always passage cells when they reach 80–90% confluency.
2. RNA is unstable and degrades rapidly. Use clean gloves and wipe working area with RNase inhibitor solution prior to RNA extraction. Always keep RNA samples on ice when thawing and preparing reagents.
3. To avoid variation in qPCR results, ensure good pipetting techniques, homogenous master mix solution, and accurate measurement of RNA concentration for cDNA synthesis.
4. Consider using codon optimization tools (e.g., IDT, GenScript, etc.) to enhance Cas13 expression and editing efficiency.
5. Inject slowly over 20 seconds for gentle release of drug without elevating eye pressure.
6. Image animal eyes during FFA promptly to ensure fluorescein signal is optimal. Longer times can affect image quality due to strong fluorescence.

Acknowledgments

This work was supported by grants from the National Health and Medical Research Council of Australia (1185600; GSL). The Centre for Eye Research Australia receives Operational Infrastructure Support from the Victorian Government.

Conflicts of Interest The authors have no conflicts of interest to declare that are relevant to the content of this chapter.

References

1. Yang S, Zhang J, Chen L (2020) The cells involved in the pathological process of diabetic retinopathy. Biomed Pharmacother 132: 110818. https://doi.org/10.1016/j.biopha.2020.110818
2. Wang J-H, Roberts GE, Liu G-S (2020) Updates on gene therapy for diabetic retinopathy. Curr Diab Rep 20(7):22. https://doi.org/10.1007/s11892-020-01308-w
3. Jinek M, Chylinski K, Fonfara I, Hauer M, Doudna JA, Charpentier E (2012) A programmable dual-RNA–guided DNA endonuclease in adaptive bacterial immunity. Science 337(6096): 816–821. https://doi.org/10.1126/science.1225829
4. Abudayyeh OO, Gootenberg JS, Konermann S, Joung J, Slaymaker IM, Cox DBT, Shmakov S, Makarova KS, Semenova E, Minakhin L, Severinov K, Regev A, Lander ES, Koonin EV, Zhang F (2016) C2c2 is a single-component programmable RNA-guided RNA-targeting CRISPR effector. Science 353 (6299):aaf5573. https://doi.org/10.1126/science.aaf5573
5. Konermann S, Lotfy P, Brideau NJ, Oki J, Shokhirev MN, Hsu PD (2018) Transcriptome engineering with RNA-targeting type VI-D CRISPR effectors. Cell 173(3):665–676.e614. https://doi.org/10.1016/j.cell.2018.02.033
6. van Eeden PE, Tee LBG, Lukehurst S, Lai C-M, Rakoczy EP, Beazley LD, Dunlop SA (2006) Early vascular and neuronal changes in a VEGF transgenic mouse model of retinal neovascularization. Invest Ophthalmol Vis Sci 47(10): 4638–4645. https://doi.org/10.1167/iovs.06-0251
7. Chuang Y-F, Wang P-Y, Kumar S, Lama S, Lin F-L, Liu G-S (2021) Methods for in vitro CRISPR/CasRx-mediated RNA editing. Front Cell Dev Biol 9:1388. https://doi.org/10.3389/fcell.2021.667879
8. Ramachandran PS, Lee V, Wei Z, Song JY, Casal G, Cronin T, Willett K, Huckfeldt R, Morgan JIW, Aleman TS, Maguire AM, Bennett J (2016) Evaluation of dose and safety of AAV7m8 and AAV8BP2 in the non-human primate retina. Hum Gene Ther 28(2):154–167. https://doi.org/10.1089/hum.2016.111
9. Zudaire E, Gambardella L, Kurcz C, Vermeren S (2011) A computational tool for quantitative analysis of vascular networks. PLoS One 6(11): e27385. https://doi.org/10.1371/journal.pone.0027385

Part II

Assays for Retinal Gene Therapy Validation In Vitro Including Patient iPSCs Generation, Retinal Organoids, and Explants

Chapter 3

Optimized Retinal Cell Dissociation for Downstream Analysis of Cone Photoreceptors

Alicia A. Brunet, Paula I. Fuller-Carter, and Livia S. Carvalho

Abstract

The isolation of viable and homogenous retinal cells is essential for downstream molecular analysis, such as single-cell sequencing and transcriptomics. However, the dissociation of retinal cells can be challenging due to the intricate nature of retinal tissue and the fragility of its cells. Furthermore, cone photoreceptors are highly susceptible to degeneration. We developed an optimized protocol for retinal cell dissociation that improves the quality of cones for downstream molecular analysis. Our protocol employs a papain enzymatic digestion, mechanical dissociation, and filtration to obtain a high percentage of viable cells with minimal cell death and damage. This protocol is a reliable and efficient method for downstream molecular analysis of cones and other retinal cells which can be useful for disease modeling—in which target cells are already sensitive and in a degenerative state—and examining the effects of treatments such as gene replacement therapies.

Key words Cones, Retina, Neurobiology, Cell dissociation, Papain, Transcriptomics, Disease modeling

1 Introduction

The retina is a complex and highly organized tissue that contains various types of specialized cells, including photoreceptors, which are crucial for visual perception [1]. The study of retinal cells—particularly photoreceptors—at the molecular level can provide important insights into the mechanisms of retinal degeneration in which the photoreceptors are primarily affected, such as age-related macular degeneration and inherited retinal diseases. However, the isolation of viable and homogenous retinal cells is a challenging task due to the intricate nature of the retina and the fragility of its cells [2–4]. One of the major challenges in dissociating retinal cells is maintaining the structural integrity of photoreceptors, especially cone photoreceptors, which are highly sensitive to mechanical and enzymatic stress. The dissociation process can lead to cell death and damage, resulting in the loss of RNA integrity and leakage of

Livia S. Carvalho (ed.), *Retinal Gene Therapy: Methods and Protocols*, Methods in Molecular Biology, vol. 3023,
https://doi.org/10.1007/978-1-0716-5198-8_3,
© The Author(s), under exclusive license to Springer Science+Business Media, LLC, part of Springer Nature 2026

intracellular RNA into the extracellular space, which can affect downstream molecular analysis and add contaminants from other cell types. Particularly, under diseased conditions, cones are even more susceptible to death making dissociation of viable cells difficult [4, 5]. Therefore, a number of issues arise when attempting to obtain a single-cell suspension of retinal tissue for further processing and analysis.

To date, various methods have been developed for retinal cell dissociation. Enzymatic digestion is commonly used for retinal cell dissociation, with papain being one of the most frequently used enzymes. Papain enzymatic digestion has been shown to effectively dissociate retinal cells and is a superior enzyme as compared to others [4–6]. However, the over-digestion of cells can lead to low cell yield and poor RNA quality, meaning a small level of mechanical dissociation is required in conjunction with enzymatic dissociation [7, 8]. Mechanical dissociation is breaking down tissue using physical force, such as cutting, crushing, or pipetting tissue and collecting loose cells. Mechanical dissociation must be done sparingly as it can lead to a high level of cell death and damage [8]. Additionally, filtration of dissociated cells can remove debris and unwanted cell aggregates, but it can also cause further cell damage if the filtration process is not optimized [9]. Therefore, there is a need for a standardized protocol that incorporates all these techniques to obtain the highest quality of cell samples and limit cell damage and death, particularly in cones.

In this study, we developed such a protocol that can provide a reliable and efficient retinal dissociation method for downstream molecular analysis of cone photoreceptors that limits cone cell death (Fig. 1).

2 Materials

2.1 Solutions

1. Digestion solution: Earle's balanced salt solution (EBSS) to make up solution to 500 μL/sample, papain 12.5 units/sample (*see* **Note 1**), 1 mM L-cysteine, 0.5 mM EDTA, DNase I 100 units/sample (*see* **Note 2**), 10% trehalose (*see* **Note 3**).
2. Inactivation solution: EBSS to make up solution to 1000 μL/sample, 20 mg ovomucoid protease inhibitor (OMI), DNase I 50 units/sample, 10% trehalose, 1% fetal calf serum (FCS; *see* **Note 4**), 0.4% bovine serum albumin (BSA; *see* **Note 5**).
3. Resuspension solution: Hank's balanced salt solution (HBSS [Ca^{2+}, Mg^{2+}, no phenol red]) to make up solution to 500 μL/sample, 10% FCS (*see* **Note 6**), 0.4% BSA.

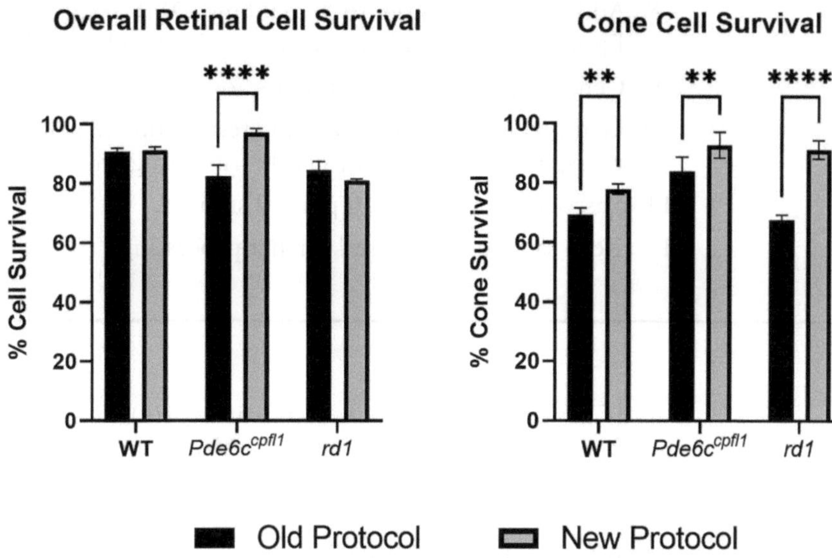

Fig. 1 Improved cone cell survival in optimized retinal dissociation protocol. Mice used included wild-type (WT) mice bred on a C57BL/6 J background, a primary cone degeneration model the cone photoreceptor functional loss 1 (*Pde6c*^{cpfl1}) in early stages of cone degeneration, and secondary cone degeneration model the retinal degeneration 1 (*rd1*) during peak cone degeneration. Retina was dissociated, and flow cytometry was performed to identify cones and dead cells. Cell survival was only significantly increased in the *Pde6c*^{cpfl1} with the new dissociation protocol ($P < 0.0001$). Cone cell survival was significantly increased in all mouse models with the new dissociation protocol (**$P < 0.01$, ****$P < 0.0001$)

2.2 Equipment

1. 1.5 mL Eppendorf tubes (collection tubes).
2. 2 mL Eppendorf tubes.
3. Fire polished glass Pasteur pipette for mechanical dissociation (*see* **Note 7**).
4. 40 μm cell strainer.

3 Methods

1. Preheat digestion solution for 30 min at 37 °C to dissolve and activate papain.
2. Collect eyes in collection tubes with HBSS on ice (*see* **Note 8**).
3. Dissect retina out and place into new collection tubes with HBSS and keep on ice.
4. Transfer retina into warm digestion solution (*see* **Note 9**).
5. Incubate at 37 °C for up to 15 min (*see* **Note 10**).
6. Remove digestion solution from each retina and immediately add 500 μL of room-temperature inactivation solution.
7. Gently triturate retina with glass pipette until dissociated. Do not triturate past the point of no clumps as over-trituration can cause cell damage and death.

8. Add cell suspension to remaining OMI solution in 2 mL Eppendorf tube and gently mix (*see* **Note 11**).
9. Centrifuge sample at 100 rcf for 2 min.
10. Gently pour off supernatant and resuspend cell pellet in 500 uL resuspension solution by gently triturating with glass pipette.
11. Pass sample through 40 μm cell strainer.
12. Keep sample on ice until further processing.

4 Notes

1. If papain is lyophilized, reconstitute 100 units of papain in 2 mL of EBSS. Some suppliers advise against storing reconstituted papain at −20 °C, but we have found no difference in frozen compared to freshly reconstituted papain.
2. If DNase I is lyophilized, reconstitute 1000 units of DNase I in 2 mL of EBSS. When reconstituting DNase I, mix gently as DNase I is sensitive to shear denaturation.
3. Trehalose was added as it can protect against desiccation and oxidative stress by preserving cell membranes and proteins from damage and denaturation [10].
4. FCS is rich in proteins and nutrients that help cell survival [11].
5. BSA has been shown to promote cell survival and is an addition to the original dissociation protocol [12].
6. We use an increased amount of FCS in the resuspension solution compared to the inactivation solution to facilitate cell survival as this is the final solution that cells will be in before further processing. For example, if processing multiple samples for fluorescent activated cell sorting, samples are being left on ice for extended amounts of time, leading to decreased number of viable cells.
7. Fire polished glass Pasteur pipettes are used for trituration to decrease stress on the sample by decreasing resistance to solution flow [4].
8. Ensure the HBSS is ice-cold to preserve the eyes and retina.
9. While the digestion solution is incubating for 30 min at 37 °C, prepare retinal samples and time the finish of preparation with the end of incubation of the digestion solution. Once samples are prepared, place it straight into the digestion solution. If incubation of the digestion solution finishes before sample preparation, the digestion solution can be left at room temperature for a short amount of time, but avoid leaving for extended amounts of time.

10. The amount of time the retina is incubated at 37 °C depends on the sample. For example, severely degenerated retinas will need less incubation time since the retina is not as thick. We have found that severely degenerated retinas can incubate for as little as 7 min.
11. 2 mL Eppendorf tubes are used before the centrifuge step as these provide a larger surface area at the bottom so the cells are less likely to be pelleted on top of each other.

Conflicts of Interest

The authors have no conflicts of interest to declare that are relevant to the content of this chapter.

References

1. Kolb H (2003) How the retina works: much of the construction of an image takes place in the retina itself through the use of specialized neural circuits. Am Sci 91(1):28–35
2. Wahlin KJ, Lim L, Grice EA, Campochiaro PA, Zack DJ, Adler R (2004) Technical Brief A method for analysis of gene expression in isolated mouse photoreceptor and Müller cells. Mol Vis 10:366–375
3. Zayas-Santiago A, Derwent JJK (2009) Preservation of intact adult rat photoreceptors in vitro: study of dissociation techniques and the effect of light. Mol Vis 15:1
4. Fadl BR, Brodie SA, Malasky M, Boland JF, Kelly MC, Kelley MW et al (2020) An optimized protocol for retina single-cell RNA sequencing. Mol Vis 26:705
5. Voigt AP, Mullin NK, Stone EM, Tucker BA, Scheetz TE, Mullins RF (2021) Single-cell RNA sequencing in vision research: insights into human retinal health and disease. Prog Retin Eye Res 83:100934
6. Brunet AA, Fuller-Carter PI, Miller AL, Voigt V, Vasiliou S, Rashwan R et al (2020) Validating fluorescent Chrnb4. EGFP mouse models for the study of cone photoreceptor degeneration. Transl Vis Sci Technol 9(9): 28–28
7. Reichard A, Asosingh K (2019) Best practices for preparing a single cell suspension from solid tissues for flow cytometry. Cytometry A 95(2): 219–226
8. Jager LD, Canda C-MA, Hall CA, Heilingoetter CL, Huynh J, Kwok SS et al (2016) Effect of enzymatic and mechanical methods of dissociation on neural progenitor cells derived from induced pluripotent stem cells. Adv Med Sci 61(1):78–84
9. Cunha B, Peixoto C, Silva MM, Carrondo MJ, Serra M, Alves PM (2015) Filtration methodologies for the clarification and concentration of human mesenchymal stem cells. J Membr Sci 478:117–129
10. Luyckx J, Baudouin C (2011) Trehalose: an intriguing disaccharide with potential for medical application in ophthalmology. Clin Ophthalmol 5:577–581
11. Hedlund TE, Miller GJ (1994) A serum-free defined medium capable of supporting growth of four established human prostatic carcinoma cell lines. Prostate 24(5):221–228
12. Francis GL (2010) Albumin and mammalian cell culture: implications for biotechnology applications. Cytotechnology 62(1):1–16

Chapter 4

Development of In Vitro Potency Assays for Gene Therapy Products Targeting cGMP-Related Retinopathies

Marina Pavlou and Stylianos Michalakis

Abstract

The emergence of efficient viral vectors derived from adeno-associated viruses (AAV) has led many groups to develop gene therapies for inherited retinal diseases. To evaluate the potency of new gene therapy vectors in a preclinical context, it is common to use animal models, such as gene-deficient or mutant animal models of a given human disease, and then assess transgene expression and resulting restoration of vision with functional or behavioral assays. While such animal models are invaluable to the preclinical testing process, they cannot be readily used as batch release tests during manufacturing or to validate biological activity at later stages of development. As such, there is need for rapid and reliable in vitro models to test whether therapeutic vectors can deliver their genetic payload and, more importantly, whether this leads to the intended biological function. We developed a cell-based system to evaluate AAV vectors designed to deliver a healthy *CNGA3* gene copy into photoreceptors of patients with *CNGA3*-linked achromatopsia. Our system uses immortalized cells that express a rhodopsin guanylyl cyclase (RhGC) from *Blastocladiella emersonii* and a genetically encoded calcium indicator (GECI) in a tetracycline-inducible manner. When transduced cells express a functional AAV-borne CNG channel, the three proteins generate a fluorometric assay that relies on the interdependent activity of its components. Using this system, we were able to confirm and quantify the function of the ion channel encoded by AAV/CNGA3 and differentiate between AAV vector potencies. Finally, we show that this approach can be readily adapted for the assessment of AAV-borne phosphodiesterase function.

Key words Retina, AAV, cGMP, In vitro potency assay, Biological activity, Transgene expression

1 Introduction

Monogenic retinal diseases are ideal candidates for gene therapy as they are both genetically and phenotypically well characterized. As such, gene therapy that supplements a single gene has been shown to provide therapeutic benefit in autosomal recessive inherited diseases. The emergence of recombinant adeno-associated viral (AAV) vectors, with tropism for retinal cells, has made it possible to locally deliver gene therapies with intraocular injections. This has sparked numerous efforts to develop new and optimized AAV-based gene

Livia S. Carvalho (ed.), *Retinal Gene Therapy: Methods and Protocols*, Methods in Molecular Biology, vol. 3023,
https://doi.org/10.1007/978-1-0716-5198-8_4,
© The Author(s), under exclusive license to Springer Science+Business Media, LLC, part of Springer Nature 2026

therapy products to treat the plethora of blinding monogenic diseases. Throughout the development of new gene therapies, there is need for informative in vitro assays that can be performed rapidly and cost-effectively before in vivo testing to facilitate the evaluation of different promoters, codon-optimized genes, and optimized regulatory genetic elements. In addition, functional potency assays are required for quality control and batch release testing during good manufacturing practice (GMP) production and for stability studies of a final gene therapy drug product.

A monogenic disorder for which five gene therapy candidates are presently being tested in clinical trials is achromatopsia. Achromatopsia is characterized by severe impairment of visual acuity and color vision as well as photophobia and pendular nystagmus. To date, six genes (*ATF6A, CNGA3, CNGB3, GNAT2, PDE6C,* and *PDE6H*) are known to cause achromatopsia, with mutations in *CNGA3* and *CNGB3* responsible for more than 90% of cases [1]. *CNGA3* and *CNGB3* encode the alpha and beta subunits of the cone cyclic nucleotide-gated (CNG) channel that mediates the depolarizing Na^{+}/Ca^{2+} "dark" current in cone photoreceptor outer segments [2]. Previously, the potency of gene therapy vectors for the treatment of *CNGA3*- or *CNGB3*-linked achromatopsia was assessed after subretinal or intravitreal injection of the gene therapy vector in gene-deficient animal models (i.e., mouse or dog) [3–10]. After an incubation period of several weeks to months, the treatment effect was determined by gene expression analyses to assess *CNGA3* or *CNGB3* transgene expression and vision-guided behavior and/or electroretinography (ERG) to indirectly assess the biological activity of the expressed CNGA3/CNGB3 cGMP-gated ion channel. Such in vivo potency testing of gene therapy vectors has low throughput and can only be performed by trained personnel in specialized animal facilities equipped for ophthalmological procedures and examinations. This also depends on the availability of a sufficient number of animals and can take several weeks or months to establish therapeutic efficacy. As such, a rapid and simple in vitro system that could functionally evaluate new viral vectors would substantially streamline potency testing of gene therapy vectors while reducing costs.

To this end, we engineered a cell-based system that can indirectly track the presence of a functional AAV-borne CNG channel using a simple fluorometric assay. This assay uses a light-inducible rhodopsin guanylyl cyclase (RhGC) from *Blastocladiella emersonii* [11] to produce cGMP following illumination at 530–560 nm and a genetically encoded calcium indicator (GECI) that is sensitive to changes in calcium content at high-throughput capacity [12, 13] (Fig. 1). We show that this system can be used to reliably assess the biological function of AAV vectors expressing CNGA3 and, with some adaptation, vectors expressing PDE6A [14] (Fig. 2).

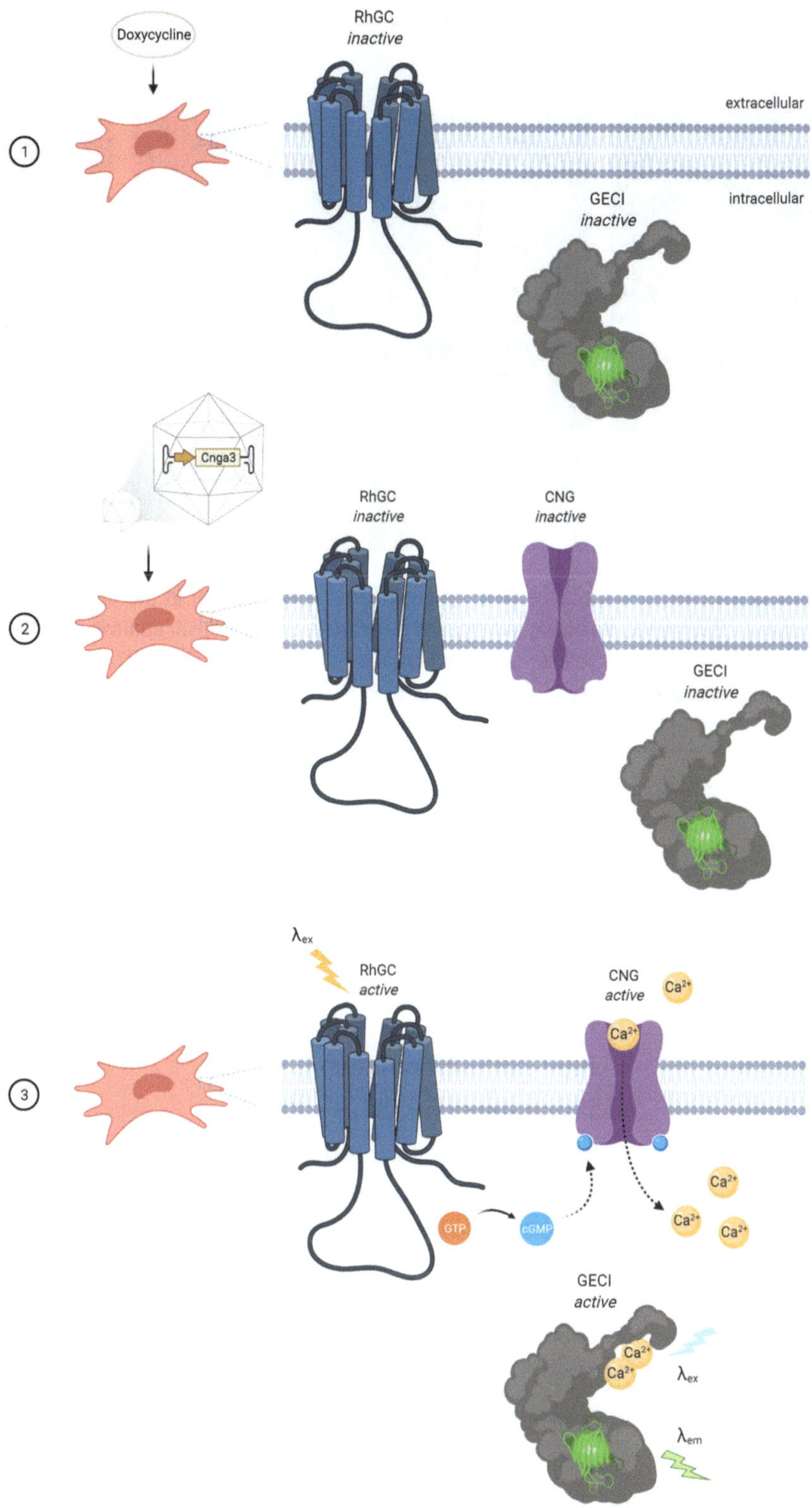

Fig. 1 Schematic representation of the in vitro platform where (1) doxycycline induces the expression of a genetically encoded calcium indicator and the rhodopsin guanylyl cyclase, (2) AAV-mediated delivery of a cGMP-gated channel subunit leads to the respective protein expression and channel formation, and (3) illumination of the rhodopsin guanylyl cyclase produces cGMP that binds to the cGMP-gated channel and permits calcium ion influx that then illuminates the calcium indicator

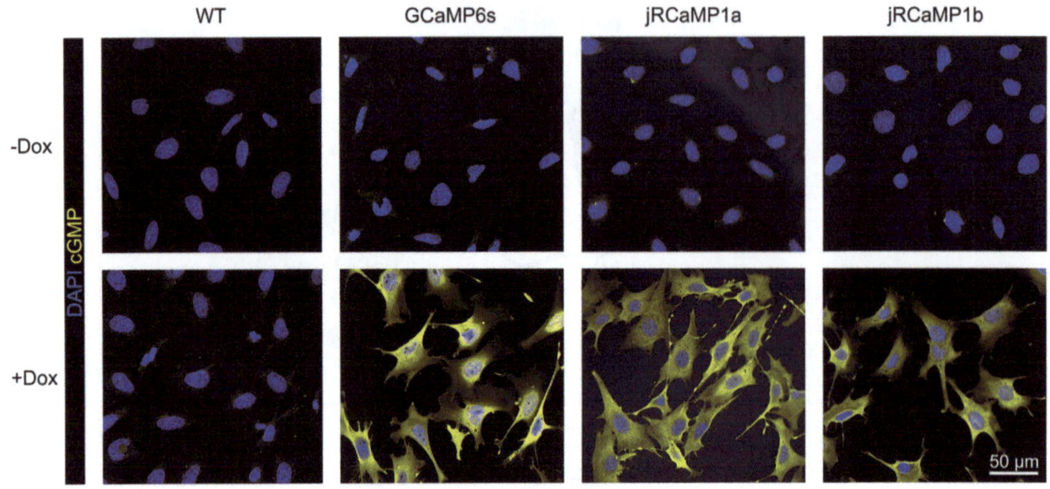

Fig. 2 Representative fluorescent images of three engineered cell lines that in the presence of doxycycline express detectable cGMP levels following illumination to activate the rhodopsin guanylyl cyclase. DAPI nuclei shown in blue, cGMP shown in yellow; WT: wild-type cells; GCaMP6s, jRCaMP1a, jRCaMP1b: calcium-indicator cell lines; Dox: doxycycline; scale bar: 50 um

2 Materials

2.1 Solutions

1. Culturing medium: low glucose DMEM, 10% fetal bovine serum (FBS), 1% Anti/Anti.
2. Culturing medium with puromycin: low glucose DMEM, 10% FBS, 1% Anti/Anti, 1 mg/mL puromycin.
3. Culturing medium with doxycycline: low glucose DMEM, 10% FBS, 1% Anti/Anti, 0.5 µg/mL doxycycline.
4. Permeabilization solution: $HBSS^{+CaCl}2$, 8 µM ionomycin.
5. Fixing solution: PBS, 4% PFA.
6. Nuclear staining solution: PBS, 1:2000 DAPI.
7. Blocking solution: PBS, 2% bovine serum albumin (BSA), 0.3% Triton-X.

2.2 Consumables

1. 1× PBS2. $DPBS^{-CaCl}{}_2$.
2. $HBSS^{+CaCl}{}_2$.
3. DAPI.
4. Ionomycin.
5. Doxycycline.
6. Puromycin.
7. DMSO.
8. 0.05% Trypsin/EDTA.
9. 0.4% Trypan Blue.

10. Thermosensitive alkaline phosphatase enzyme (1 U/μL).
11. Sterile glass coverslips.
12. Poly-D-lysine.
13. Reverse transcriptase kit.
14. SYBR Green.
15. RNase-free DNase kit.
16. Gel extraction kit.
17. Genome-integrating vector plasmid, e.g., piggyBac.
18. Human cells, e.g., HeLa.
19. Flat-bottom 24-well cell culture plates.

2.3 Equipment

1. Any commercially available microplate reader such as SpectraMax plate reader Multi-Mode Microplate Reader from Molecular Devices or equivalent. Suggested settings: high photomultiplier tube (PMT) and optics acquisition settings for maximum sensitivity.
2. Any commercially available thermal cycler that can perform PCR.
3. Any commercially available real-time PCR platform.

3 Methods

Prepare cell culture reagents under aseptic conditions and molecular biology reagents under clean laboratory conditions. Warm up all cell culture reagents to room temperature before use. Follow standard cell culture practices for immortalized HeLa cells and AAV vector production.

All illumination and fluorescence recordings are performed using the microplate reader. Set the plate reader temperature to 37 °C before all recordings and load the microinjector with an 8 μM ionomycin solution in $HBSS^{+CaCl}{}_2$

3.1 Cell Culture

1. Prepare complete culturing medium using low glucose DMEM supplemented with 10% FBS and 1% Anti/Anti.
2. To bring cryopreserved cells to a culture, first thaw cryovials and immediately resuspend cell suspension in warmed medium.
3. Centrifuge cell suspension at 0.5 rcf for 5 min and remove supernatant containing DMSO. Then resuspend the cell pellet in complete medium and transfer an appropriate cell suspension volume to a sterile T-75 flask.
4. Cells are maintained in an incubator at 37 °C, 5% $^{CO}{}_2$, 85–95% humidity, and passaged once 90% confluent.

5. To passage the cells, aspirate the medium and wash once with DPBS$^{-CaCl_2}$. Then detach cells from the flask using 0.05% Trypsin/EDTA (5′ incubation at 37°C, 5% CO_2).
6. To quench the digestion, add 3 volumes of complete medium and transfer the cell suspension to a falcon tube.
7. Count cells using 0.4% Trypan Blue Stain and an automated cell counter before seeding the desired cell number into a new recipient. For a 24-well plate, seed 50,000 cells/well.
8. To transfect cells, use an appropriate protocol for the plasmid size you want taken up by the cells.

3.2 Generation of New HeLa Cell Lines

1. Use a vector plasmid system, e.g., piggyBac, that allows you to integrate a desired gene cassette into the genome of HeLa cells to create new cell lines (*see* **Notes 1–3**).
2. Choose a vector plasmid that has an antibiotic resistance gene that allows for colony selection in eukaryotic cells, e.g., puromycin, and for inducible expression of your cassette, e.g., tetracycline-inducible gene expression (*see* **Notes 1–3**).
3. Use NEBuilder (https://nebuilder.neb.com) to determine the optimal cloning strategy for your constituents. To enable the simultaneous expression of multiple genes under the same conditions, generate a polycistronic cassette that is linked either by 2A nucleotides or an IRES.
4. Linearize the vector plasmid via enzymatic digestion followed by a thermosensitive alkaline phosphatase enzyme (1 U/μL) treatment to prevent spontaneous reannealing.
5. Perform PCR to amplify the coding sequences of RhGC and GECI genes using the primers generated from the NEBuilder cloning strategy (*see* **Note 2**).
6. Extract the linear DNA fragments following agarose gel electrophoresis using a gel extraction kit and perform Gibson Assembly using the instructions from NEBuilder (*see* **Note 3**).
7. Transfect HeLa cells with the resulting vector plasmid (s) required for genome integration, i.e., piggyBac vector plasmid and transposase plasmid. To select for cells that had successful genome integration, apply antibiotic, e.g., puromycin (1 mg/mL), for 7 days.
8. To select a single colony of HeLa cells, passage cells as usual and then perform a series of consecutive dilutions until you have a cell suspension of approximately 1.5 cells/mL.
9. Seed 1 mL of suspension into each well of a 96-well plate and monitor cell growth until confluence. Passage cells gradually into larger containers until you have a consistent colony of engineered HeLa cells.

10. Extract gDNA from the single colonies and use 200 ng as template to perform quantitative PCR (qPCR) of your gene cassette to determine which colony had the highest number of inserts and use that for subsequent experiments.

3.3 AAV Vector Production and Cell Transduction

1. Use single-stranded AAV plasmids containing a mSWS-CNGA3, hArr3-CNGA3, and hRho-PDE6A expression cassette flanked by ITRs from AAV2 as vector plasmids.
2. Produce AAV vectors using an established protocol as previously described [9] or purchase from a commercial company.
3. Determine genomic titers of AAV preps by qPCR using primers: ITR2-F: 5′ ggaacccctagtgatggagtt 3′ and ITR2-R: 5′ cggcctcagtgagcga 3′.
4. Test the multiplicity of infection (MOI) that results in optimal transduction efficiency in your cells. To this end, calculate the volume of AAV solution required to expose each cell to the desired number of vector particles and add the required vector volume in the media over the cells; incubate for 48 h and assess transduction efficiency using, e.g., a fluorescent reporter.
5. For a test experiment, seed cells (50,000 cells/well) in a clear flat-bottom 24-well plate and culture in complete medium supplemented with doxycycline (0.5 μg/mL) to induce the expression of RhGC and GECI in the engineered cells. Once attached (~4 h), transduce cells with AAVs carrying *CNGA3* at the optimal MOI by adding the appropriate vector volume in the media and incubate for 48 h.

3.4 Gene Expression Analysis

1. Extract RNA from AAV-transduced cells and digested with an RNase-free DNase kit to eliminate DNA contaminants.
2. Reverse transcribe 50–200 ng of RNA to cDNA using a reverse transcriptase kit such as SuperScript IV Reverse Transcriptase (Thermo Fisher) or RevertAid First Strand cDNA Synthesis Kit (Thermo Fisher).
3. Use primer pairs for eGFP, ITR2, and GECIs (*see* Table 1) and perform quantitative PCR by using SYBR Green according to manufacturer instruction on any commercially available real-time PCR platform.

3.5 Immunocytochemistry and Imaging

1. Seed cells on sterile glass coverslips that were previously coated with poly-D-lysine according to the manufacturer instructions.
2. To verify light-inducible cGMP production, illuminate the attached cells at the excitation wavelength (λex) of RhGC, i.e., 500–560 nm in 10 nm intervals, and then immediately fix with 4% PFA/PBS for 10 min at room temperature.

Table 1
Sequences of primers used for quantitative PCR

Target	Forward primer	Reverse primer	Annealing T°C
eGFP	CGACCAC TACCAGCAGAACAC	TTCTCGTTGGGGTCTTTGC TCAG	62
ITR2	GGAACCCCTAGTGATGGAG TT	CGGCCTCAGTGAGCGA	58
GcaMP6s	CCCGACAACCACTACCTGAG	GTCCATGCCGAGAGTGATCC	60
jRCaMP1a	AAGACAGGTCACGCAGTCAG	GAGTGTAACCACGCAGACCA	60
jRCaMP1b	GGAATAAGTGGGGTCACGCA	GAGTGTAACCACGCAGACCA	60

3. For staining, wash the cell monolayer twice with 1× PBS and then incubate overnight at 4 °C with primary antibodies diluted in blocking solution (2% BSA/0.3% Triton-X in PBS) (*see* **Note 4**).
4. The next day wash the cell monolayer twice with 1× PBS and incubate with a secondary antibody solution in 1× PBS for 1.5 h at room temperature while covered (*see* **Note 5**).
5. Finally, wash the monolayer once with 1× PBS and incubate for 10 min with DAPI 1:2000 prior to mounting with a glass slide.
6. Acquire fluorescent images with an inverted confocal microscope. You can further process the original images, consisting of multiple z-stacks, with the open-source software such as Fiji.

3.6 RhGC Illumination and GECI Recordings

1. Perform all recordings using clear flat-bottom 24-well plates and read the signal from the bottom of the well with a high photomultiplier tube (PMT) and optics setting.
2. To account for discrepancies of cell distribution in each well, record from 12 dispersed points from each well and then average the signal obtained.
3. Perform six technical repeats for each condition to account for transduction variability.
4. Ahead of GECI recordings in the plate reader, replace the media with DPBS-$^{CaCl}{}_2$ and wash twice to ensure residual phenol red from the media is removed.
5. Place the 24-well plate in the pre-warmed (37 °C) plate reader and illuminate first with a 500–560 nm spectrum at 10 nm intervals to activate RhGC.
6. Then record a baseline GECI signal with single wavelength settings. For GcaMP6s: λex 488—λem 535 nm and for jRCaMP1a-b: λex 558—λem 605 nm (*see* **Note 6**).

7. Subsequently, replace $DPBS^{-CaCl_2}$ with $HBSS^{+CaCl_2}$ to test the GECI signal using the same wavelength settings as before. This allows you to make paired comparisons between any baseline signal and test signal (*see* **Note 7**).
8. For the final step, remove most of the $HBSS^{+CaCl_2}$ and leave only 100 μL covering the cells. Reinsert the plate in the plate reader, and using the SmartInject setting, inject 100 μL of 8 μM ionomycin (*see* **Note 7**) over the cells, use the plate reader setting to shake the plate to mix (final concentration 4 μM), and after a 5-s pause record the control GECI signal (*see* **Note 8**).

4 Notes

1. We used a piggyBac vector with backbone: pB09/TRE-MCS-SV40pA-Ef1a-rtTA-PuroR- SV40pA (gift from Prof. Volker Busskamp, University of Bonn). PiggyBac vector systems allow seamless integration of transposable elements at TTAA sites without genomic scarring. The vector backbone contains 5′ and 3′ terminal inverted repeats that flank the transgene cassette which will be integrated into the host cell genome in the presence of the piggyBac transposase [15]. A puromycin resistance gene was included in the vector backbone to select for cells that had the transgene cassette successfully integrated into their genome.
2. Template plasmids containing the coding sequence for RhGC and GECI: RhGC(hBE)_pGEM (kindly provided by Prof. Peter Hegemann, Humboldt University of Berlin), pGP-CMV-GcaMP6s (Addgene #40753), pGP-CMV-NES-jRCaMP1a (Addgene #61562), and pGP-CMV-NES-jRCaMP1b (Addgene #63136).
3. In order to regulate the expression of RhGC and GECI in a doxycycline-inducible manner, a Tet-On system was used [16]. Both genes were downstream a tetracycline response element (TRE) promoter, and a ubiquitous Ef1a promoter drove the expression of reverse tTA (rtTA) that would block expression of RhGC and GECI in the absence of doxycycline. The resulting plasmids were pB/TRE- RhGC(BE)-T2A-GCaMP6s-SV40pA-Ef1a-rtTA-T2A-PuroR-SV40pA (11 kb), pB/TRE- RhGC(BE)-T2A-jRCaMP1a-SV40pA-Ef1a-rtTA-T2A-PuroR-SV40pA (11.033 kb), and pB/TRE-RhGC(BE)-T2A-jRCaMP1b-SV40pA-Ef1a-rtTA-T2A-PuroR-SV40pA (11.033 kb).
4. The primary antibodies used are sheep anti-cGMP 1:3000 (from Prof. Steinbusch, Maastricht University, Netherlands),

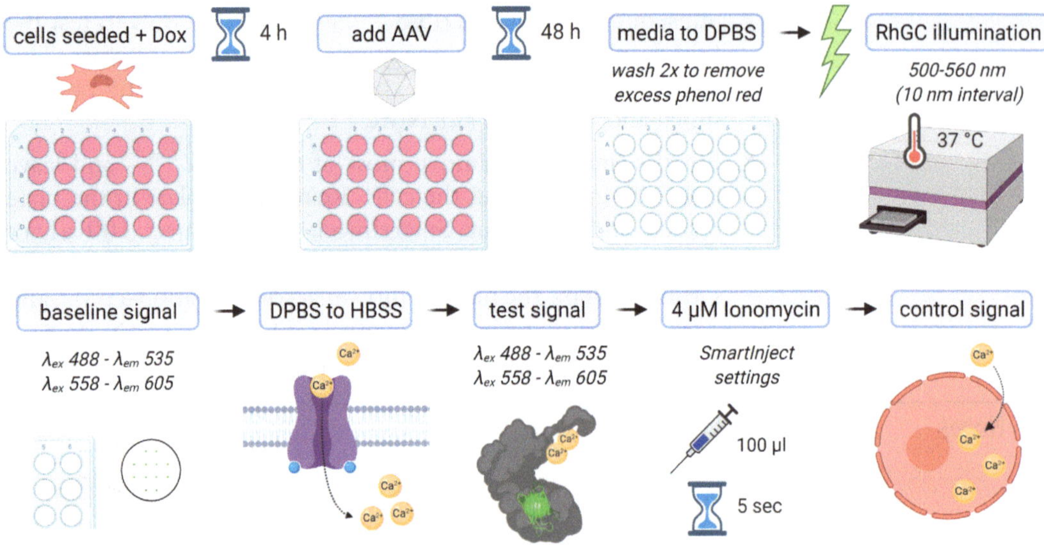

Fig. 3 Schematic representation of the method to determine functional output of the in vitro platform broken down into the consecutive steps as outlined in this protocol

rabbit anti-PDE6A IgG 1:500 (NBP1–87312, Novus Biologicals), and rat anti-CNGA3 1:30 (custom-made anti-CNGA3, 7D8, hybridoma supernatant, kind gift of Heinz G. Körschen, Caesar Bonn).

5. The secondary antibodies used are sheep anti-Alexa594 IgG (H + L), donkey 1:500 (A-11016 Thermo Fisher), anti-rabbit-Alexa488 IgG (H + L) 1:400 (#4412, Cell Signal), and anti-rat-Cy3 IgG (H + L) 1:200 (112–165-143, Jackson).
6. We used ultrasensitive genetically encoded calcium sensors fused to fluorescent proteins for noninvasive imaging. These calcium sensors were designed to detect fast calcium dynamics, optimal for neurons, and have specific excitation/emission spectra based on the fluorophore fused to the sensor. For our purposes, we tested GcAMP6s with spectra in the green channel (λex 488—λem 535 nm) and jRCaMP1a and jRCaMP1b with spectra in the red channel (λex 558—λem 605 nm) [12, 17] (Fig. 3).
7. When cells are exposed to 4 μM ionomycin, their cell membrane becomes permeabilized allowing calcium ions to freely diffuse into the cell and engage with the GECI proteins. This step should obtain the maximum GECI signal from cells and acts as a positive control that the GECI is indeed sensitive to changes in calcium content. This step will kill the cells shortly after exposure, which is why the automation of the plate reader microinjector is ideal for regulating this step.

8. Since the GECI signal is recorded from the same cells under different conditions, i.e., in the absence of extracellular calcium (DPBS$^{-CaCl_2}$) and the presence of extracellular calcium (HBSS$^{+CaCl_2}$), it allows you to perform paired comparisons of the GECI signal in your statistical analysis.

Conflict of Interest

The authors have no conflicts of interest to declare that are relevant to the content of this chapter.

References

1. Michalakis S, Gerhardt M, Rudolph G, Priglinger S, Priglinger C (2022) Achromatopsia: genetics and gene therapy. Mol Diagn Ther [Internet]. [cited 2022 Feb 16];26(1):51. Available from: /pmc/articles/PMC8766373/
2. Michalakis S, Becirovic E, Biel M (2018) Retinal cyclic nucleotide-gated channels: from pathophysiology to therapy. Int J Mol Sci [Internet]. [cited 2021 Feb 18];19(3):749. Available from: www.rcsb.org/3d-view/
3. Komáromy AM, Rowlan JS, Corr ATP, Reinstein SL, Boye SL, Cooper AE, et al (2013) Transient photoreceptor deconstruction by CNTF enhances rAAV-mediated cone functional rescue in late stage CNGB3-achromatopsia. Mol Ther [Internet]. [cited 2022 Feb 16];21(6):1131. Available from: /pmc/articles/PMC3677296/
4. Komáromy AM, Alexander JJ, Rowlan JS, Garcia MM, Chiodo VA, Kaya A, et al (2010) Gene therapy rescues cone function in congenital achromatopsia. Hum Mol Genet [Internet]. [cited 2022 Feb 16];19(13):2581–93. Available from: https://pubmed.ncbi.nlm.nih.gov/20378608/
5. Michalakis S, Mühlfriedel R, Tanimoto N, Krishnamoorthy V, Koch S, Fischer MD, et al (2010) Restoration of cone vision in the CNGA3−/− mouse model of congenital complete lack of cone photoreceptor function. Mol Ther [Internet]. [cited 2020 Mar 25];18(12): 2057–63. Available from: https://pubmed.ncbi.nlm.nih.gov/20628362/
6. Carvalho LS, Xu J, Pearson RA, Smith AJ, Bainbridge JW, Morris LM, et al (2011) Long-term and age-dependent restoration of visual function in a mouse model of CNGB3-associated achromatopsia following gene therapy. Hum Mol Genet [Internet]. [cited 2022 Feb 16];20(16):3161–75. Available from: https://pubmed.ncbi.nlm.nih.gov/21576125/
7. Pang J jing, Deng WT, Dai X, Lei B, Everhart D, Umino Y, et al (2012) AAV-mediated cone rescue in a naturally occurring mouse model of CNGA3-achromatopsia. PLoS One [Internet]. [cited 2019 Nov 15];7(4):e35250. Available from: https://journals.plos.org/plosone/article?id=10.1371/journal.pone.0035250
8. Banin E, Gootwine E, Obolensky A, Ezra-Elia R, Ejzenberg A, Zelinger L, et al (2015) Gene augmentation therapy restores retinal function and visual behavior in a sheep model of CNGA3 achromatopsia. Mol Ther [Internet]. [cited 2022 Feb 16];23(9):1423–33. Available from: https://pubmed.ncbi.nlm.nih.gov/26087757/
9. Mühlfriedel R, Tanimoto N, Schön C, Sothilingam V, Garrido MG, Beck SC et al (2017) AAV-mediated gene supplementation therapy in achromatopsia type 2: preclinical data on therapeutic time window and long-term effects. Front Neurosci 11(MAY):292
10. Pavlou M, Schön C, Occelli LM, Rossi A, Meumann N, Boyd RF et al (2021) Novel AAV capsids for intravitreal gene therapy of photoreceptor disorders. EMBO Mol Med 13:e13392
11. Scheib U, Stehfest K, Gee CE, Körschen HG, Fudim R, Oertner TG et al (2015) The rhodopsin-guanylyl cyclase of the aquatic fungus Blastocladiella emersonii enables fast optical control of cGMP signaling. Sci Signal 8(389):1–8
12. Chen TW, Wardill TJ, Sun Y, Pulver SR, Renninger SL, Baohan A, et al (2013) Ultrasensitive fluorescent proteins for imaging neuronal activity. Nature [Internet]. [cited 2023 Aug 29];499(7458):295. Available from: /pmc/articles/PMC3777791/

13. Wu N, Nishioka WK, Derecki NC, Maher MP (2019) High-throughput-compatible assays using a genetically-encoded calcium indicator. Sci Rep [Internet]. [cited 2021 Feb 19];9(1): 1–17. Available from: https://doi.org/10.1038/s41598-019-49070-8

14. Pavlou M, Babutzka S, Michalakis S (2022) A bioengineered in vitro model to assess AAV-based gene therapies for cyclic GMP-related disorders. Int J Mol Sci [Internet]. [cited 2023 May 25];23(9):4538. Available from: https://www.mdpi.com/1422-0067/23/9/4538/htm

15. Wilson MH, Coates CJ, George AL Jr (2007) PiggyBac transposon-mediated gene transfer in human cells. Mol Ther 15(1):139–45. https://doi.org/10.1038/sj.mt.6300028. PMID: 17164785

16. Das AT, Tenenbaum L, Berkhout B (2016) Tet-on systems for doxycycline-inducible gene expression. Curr Gene Ther 16(3):156–67. https://doi.org/10.2174/1566523216666160524144041. PMID: 27216914

17. Dana H, Mohar B, Sun Y, Narayan S, Gordus A, Hasseman JP, Tsegaye G, Holt GT, Hu A, Walpita D, Patel R, Macklin JJ, Bargmann CI, Ahrens MB, Schreiter ER, Jayaraman V, Looger LL, Svoboda K, Kim DS (2016) Sensitive red protein calcium indicators for imaging neural activity. Elife 5:e12727. https://doi.org/10.7554/eLife.12727. PMID: 27011354

Chapter 5

Quantification of Drusen-Like Deposits in Human Pluripotent Stem Cell-Derived Retinal Pigment Epithelium Cells Using Confocal Microscopy

Jenna Hall, Shane Cheung, Kalyan Shobhana, Ellie Cho, Grace E. Lidgerwood, Daniel Paull, and Alice Pébay

Abstract

Accessing pathologically relevant models of human retinal degeneration is crucial for understanding, screening, and developing accurate treatment strategies that prevent or slow progression of blinding diseases. Human pluripotent stem cells (hPSCs) are a powerful tool to investigate retinal disease, as cells from selected individuals or specific genetic background can be differentiated into cell types of interest. hPSC-derived retinal pigment epithelium (RPE) cells can produce drusen-like deposits that accumulate between the RPE and the underlying Bruch's membrane. These key pathological markers of specific retinal disorders are detectable using simple microscopy techniques. However, quantification of these deposits often requires slow, low-throughput manual counting of images. This further lends itself to issues including sampling biases while ignoring critical data parameters such as deposit volume and precise localization. To overcome these issues, we developed two methods for quantifying the presence of drusen-like deposits in vitro, using hPSC-derived RPE cultures. Both methods accurately quantify drusen-like deposit numbers, and the mid-throughput pipeline allows for quantification of drusen-like deposit volume, as well as discerns their apical or basal localization on RPE cells. The platform's ability to track changes in drusen-like deposits supports screening experimentation and personalized medicine initiatives, including for gene therapy optimization.

Key words Human Pluripotent stem cell, Cell culture, Retinal pigment epithelium, Confocal microscopy, Automation, Age-related macular degeneration, Quantification

1 Introduction

Various retinal dystrophies including age-related macular degeneration (AMD) are characterized by the presence of drusen, which are extracellular deposits that form between the retinal pigment epithelium (RPE) and the underlying Bruch's membrane [1]. These

Grace Lidgerwood, Daniel Paull, and Alice Pébay are equal senior authors with all other authors

Livia S. Carvalho (ed.), *Retinal Gene Therapy: Methods and Protocols*, Methods in Molecular Biology, vol. 3023, https://doi.org/10.1007/978-1-0716-5198-8_5,
© The Author(s), under exclusive license to Springer Science+Business Media, LLC, part of Springer Nature 2026

are observable clinically using fundoscopy [2]. In clinical settings, drusen deposit number and location have prognostic value in predicting disease advancement [3]; however, the availability of reliable experimental models in which to investigate progression is limited. Human pluripotent stem cells (hPSCs) and their derivatives have proven useful for studying disease, including of the retina [4]. For instance, hPSC-derived RPE cells produce drusen-like deposits, and these are detectable by microscopy [5–7]. However, methods for quantifying drusen-like deposits in vitro are manual and low throughput and often introduce bias. The ability to efficiently, reliably, and quickly quantify drusen-like deposits in vitro could open new potentials for disease modeling, drug screening, and assessment of gene therapy efficiency.

Major protein components of drusen deposits include vitronectin [8] and apolipoprotein E (APOE) [9]. Here, using antibody markers for APOE and vitronectin as proxies for drusen deposits, we describe two methods based on microscopy that allow for the quantification of drusen-like deposits in a reliable manner, amenable to either mid- or high-throughput screens [10]. These methodologies could be used for screens of candidates able to clear drusen-like deposits, select hits, and screen for impact on drusen-like deposits in patient lines for subsequent clinical translation.

These tools prove useful for retinal disease modeling and for high-throughput screening approach to assess the efficacy of potential gene therapies in reducing or preventing drusen formation, as well as to test candidate gene therapies on iPSC-derived RPE cells with disease-specific genetic mutations to identify treatments that mitigate drusen accumulation.

2 Materials

1. RPE Progenitor Specification Medium: TeSR-E6 (STEMCELL Technologies), N2 (Thermo Fisher Scientific).
2. RPE Maturation Media (RPEM): α-MEM (Life Technologies), fetal bovine serum (FBS), MEM NEAA (Life Technologies), N1 supplement (Sigma Aldrich), L-Glutamine–Penicillin–Streptomycin solution, taurine 250 μg/mL, hydrocortisone 20 ng/mL, triiodothyronine 100 ng/mL.
3. Permeabilization and blocking buffer: 0.1% Triton-X, 4% normal donkey serum, 1× $PBS^{+/+}$.
4. 4% paraformaldehyde (PFA): 50% volume 8% PFA, 40% DH2O, 10% filtered $PBS^{+/+}$.
5. Final buffer: 0.1% Triton, 100 mM glycine, 0.1% sodium azide in $PBS^{+/+}$.
6. Vitronectin (Stem Cell Technologies).
7. StemFlex (Thermo Fisher).

8. Trypsin-EDTA (0.25%) with phenol red.
9. $PBS^{-/-}$ and $PBS^{+/+}$.
10. 96-well Cell Carrier Ultra Plate.
11. Matrigel™ (Corning, coated at 0.035 mg/cm^2).
12. ZO-1 10 μg/mL.
13. APOE 3 μg/mL.
14. Vitronectin 5 μg/mL.
15. Antibody Alexa Fluor 647 4 μg/mL.
16. Antibody Alexa Fluor 488 4 μg/mL.
17. Hoechst 1 μg/mL.
18. Ms. and Rb IgG control antibodies.
19. UltraPure H_2O.
20. Confocal microscope and appropriate software such as the Zeiss LSM 900 confocal microscope and Zen Blue 3.2 software.
21. ImageJ/FIJI or equivalent image analysis software.
22. Imaris (v9.9, Oxford Instruments) or equivalent image analysis software.
23. Lie cell imaging microscope such as Operetta (Perkin Elmer) or equivalent.
24. Countess Cellometer or equivalent.

3 Methods

3.1 RPE Cell Differentiation

1. RPE cells were generated as described previously [11]; plate the iPSCs onto a vitronectin-coated six-well plate at ~200 k cells per well, and feed with StemFlex every other day.
2. After ~5–7 days of hPSC culturing, or once the wells are ~60% confluent, replace StemFlex medium with RPE Progenitor Specification Medium.
3. Change the media every second day.
4. After 29 days, change the media to RPEM maturation medium, with medium changes every second day.
5. After an additional 30 days, dissociate fully differentiated (D60) RPE using Trypsin-EDTA.
6. Spin cells at 1200 rpm for 5 min, aspirate supernatant, resuspend in 1 mL RPEM, and count on a Countess Cellometer.
7. Plate cells at a density of 0.075×10^6 cells/cm^2 onto a 96-well PhenoPlate coated in Matrigel™ (0.035 mg/cm^2).
8. Mature cells for a further 30 days (~90 days in total), with media changes every second day.

3.2 RPE Cell Immunohistochemistry

1. Gently wash adherent D90 iPSC-derived RPE cells twice with $PBS^{-/-}$.
2. Fix cells with 4% paraformaldehyde (10 min, room temperature).
3. Permeabilize and block cells with the permeabilization and blocking solution (60 min, room temperature, on rocker).
4. Incubate cells with the following antibodies (all 4 °C, overnight ideally, 2 h minimally): ZO-1 (10 μg/mL) and APOE (3 μg/mL) or vitronectin (5 μg/mL) (*see* **Note 1**).
5. Wash cells once with Permeabilization Buffer and twice with $PBS^{+/+}$.
6. Incubate cells with the isotype-specific secondary antibodies Alexa 568 and Alexa 488 (each at 4 μg/mL) (room temperature, 2 h, on rocker).
7. Wash cells trice (once Permeabilization Buffer, twice $PBS^{+/+}$) and counterstain nuclei with bisBenzimide Hoechst 33342 trihydrochloride diluted in UltraPure H_2O or DI water for 10 min at room temperature on rocker.
8. Wash the wells thrice (once Permeabilization Buffer, twice $PBS^{+/+}$) and leave in the final buffer; store at 4 °C (*see* **Note 2**).

3.3 Image Acquisition

3.3.1 Set Up the Microscope

1. Open the Zen Blue software from the desktop and load your plate onto the microscope stage.
2. To turn the brightfield light on, use a preassigned button from the software by going to the "Locate" tab in Zen and clicking on the "Transmitted" button (Fig. 1a).
3. Use the joystick to move the stage x-y axis to "A1" well position using the transmitted light as a guide.
4. Go to the Channels window under "Acquisition" tab, and activate ESID detector for a single channel, to generate a transmission image of the sample. Usually works best for the channel that uses the highest laser power. During live scanning, make sure the ESID channel can be visualized (Fig. 1b).
5. Activate the "Tiles" tool at the top of the acquisition tab (Fig. 1c).
6. Under the Multidimensional acquisition section, click on "Show viewer" to open the navigation pane, and in the tiles tab select the sample carrier with your specifications (Fig. 1d) (*see* **Note 3**).

3.3.2 Calibrate the Imaging Plate

1. Calibrate the imaging plate using the 5× objective as it allows for faster stage movement and large field of view. Click "Calibrate" under Sample Carrier and the calibration wizard will open (Fig. 2a).

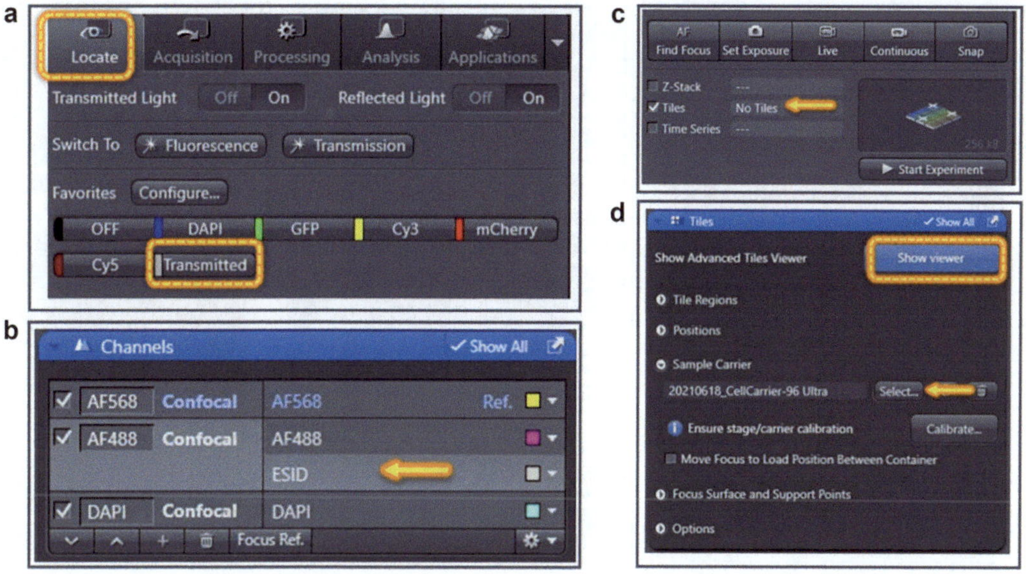

Fig. 1 Configuring LSM900 settings. Initial settings for the Zeiss LSM900 Confocal. (**a**) Under the locate tab, select the ESID or transmitted light button. (**b**) Ensure the ESID Channel is selected. (**c**) Check the Tiles tab which will open a new pane. (**d**) Select show viewer and select the plate specifications of imaging plate

2. A live view of the ESID will be displayed. If it is not, go back to the Channels tab and double-click on ESID. Click "Next" to move to the next step (Fig. 2b).
3. (Optional) In case the preview looks dim, click on "Display" under the image view area and click on Min/Max to adjust the display (Fig. 2c).
4. A window to calibrate the stage will open. Make sure the stage is calibrated on "Startup." If the stage was not calibrated, follow the instructions to calibrate the stage. Click "Next" to move to next step (Fig. 2d).
5. A window to select the calibration method will open. A list of methods based on different number points is available. Select "Search Edges of one single well (4 points)" (Fig. 2e).
6. Choose the well to search "Upper Left (A1)." This method was chosen based on time for calibration and its robustness across different sessions.
7. Click "Next" to move to the next step (Fig. 2f).
8. A window to define the edges of the wells will open (Fig. 2g).
9. Move the stage using the joystick to the left edge of the upper left well (A1). Click on "Set Position" to define the x,y position (Fig. 2h).

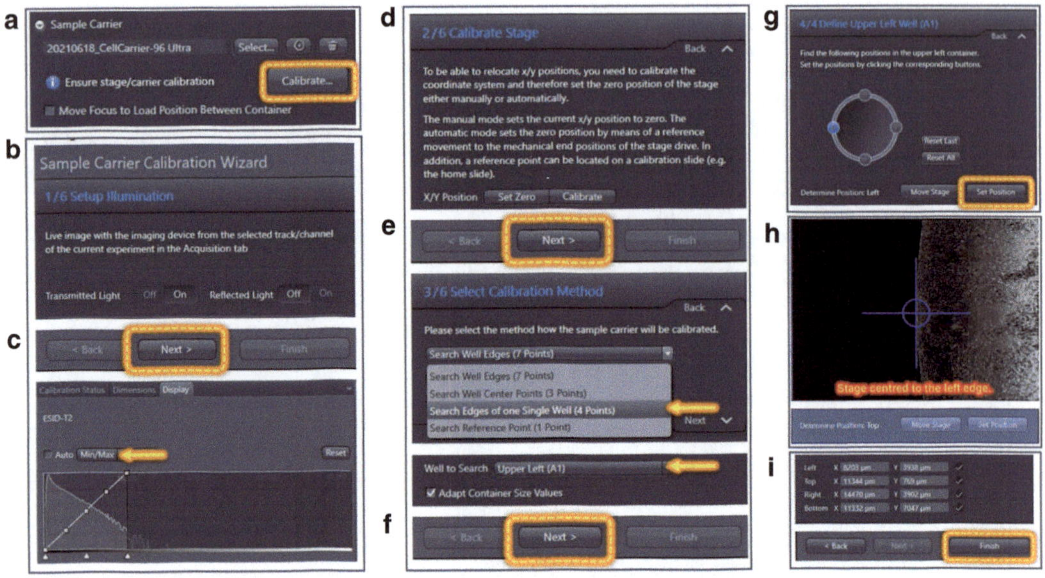

Fig. 2 Calibrate the imaging plate. Calibrate the imaging plate on the Zeiss LSM900 Confocal. (**a**) Under Sample Carrier, select calibrate. (**b**) The Calibration Wizard will open with a live view of A1. (**c**) Ensure the ESID Channel is selected; if it is and the edge of the plate is still not visible, adjust the display histogram along the bottom of the screen until it is visible. (**d**) If the stage was not calibrated on microscope startup, calibrate it before calibrating the plate. (**e**) Select Edges of one single well (4 points) as the calibration method. (**f**) Select well A1 and click next. (**g**) A window to define the edges of the wells will open. (**h**) Navigate to the leftmost center of A1, set position, and then navigate to the top, making your way around the well. (**i**) Once all four positions are set, click finish to exit out of the wizard. Plate is now calibrated

10. Click on "Move stage" to move to the top edge of the A1 well and click on "Set Position." Do some fine adjustment if the stage was not centered to the edge.
11. Repeat the above step till all the positions have been verified.
12. Click on "Finish" to exit out of the sample carrier calibration wizard (Fig. 2j).

3.3.3 Setting Up Random Position Selection

1. Change the objective to 20× (Fig. 3a).
2. Select the wells required for imaging (Fig. 3b). Note: To choose multiple wells in a series, press "Shift" and select the rows/columns. To choose selective wells, press "Ctrl" and select the specific wells.
3. Under the Positions window, click to open the options for creating position arrays for the selected carrier container (Fig. 3c).
4. Select the method to distribute positions by "Number" and input the number of points to be distributed in the wells (Fig. 3d).
5. Select the distribution as "Random."

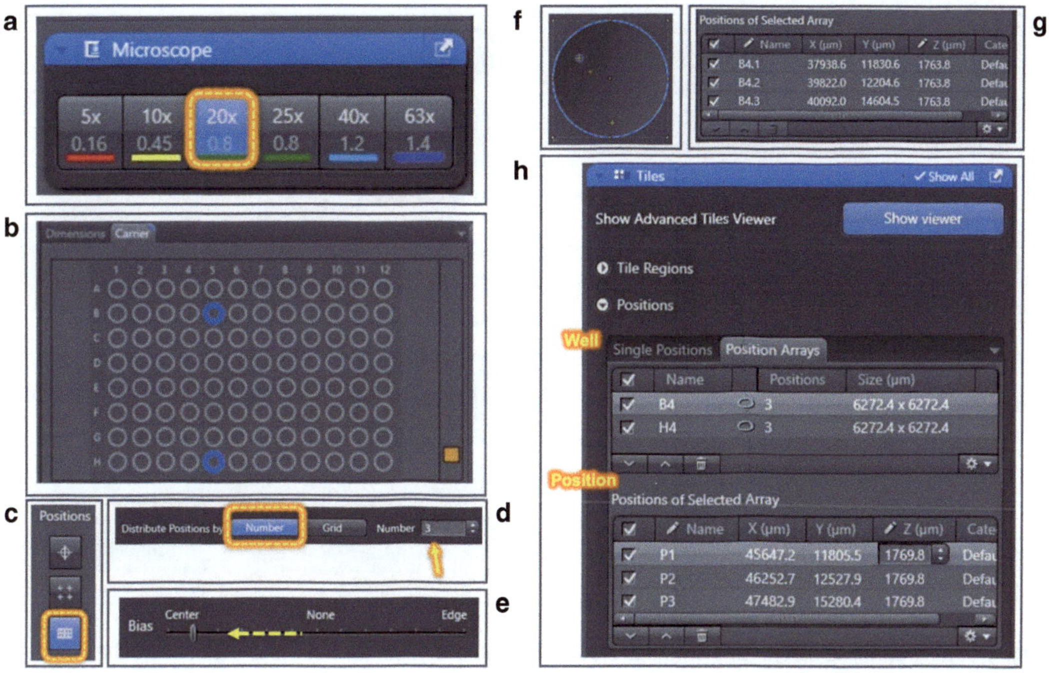

Fig. 3 Randomize image positions to reduce bias in image acquisition. (**a**) Change the objective to 20×. (**b**) Select wells of interest. (**c**) Select the position array button to open options. (**d**) Under Distribute positions, select Number, and indicate the number of random positions to be distributed per well. (**e**) Toggle the bias toward center to avoid random positions landing on the edges of wells. (**f**) Random positions will populate each selected well. (**g**) The coordinates of each random position per well will populate the Positions of Selected Array Table. (**h**) The positions per well are nested within the wells; however, all will be titled P1, P2, and P3, and it is recommended to rename them per well

6. (Optional) For image analysis, sometimes there may be a need to exclude positions close to the edge. To avoid distribution of positions on the edge, move the slider toward "Center" (Fig. 3e).
7. The random distribution of locations in each well will populate the plate map (Fig. 3f).
8. The list of positions and their coordinates can be found under the Tiles window (Fig. 3g).
9. The positions will populate two tables: the first is the position array, which is the well; this should automatically be labeled. If you click on the well, another table shows the positions within the well, called positions of selected array (Fig. 3h).

3.3.4 Setting Up the Z-Stack, Starting the Experiment, and Saving

1. Tick the Z-stack tool at the top of the acquisition tab (Fig. 4a). Z-stack window will appear under Multidimensional acquisition. Highlight and "double-click" on the positions in a well to move the stage to the position.

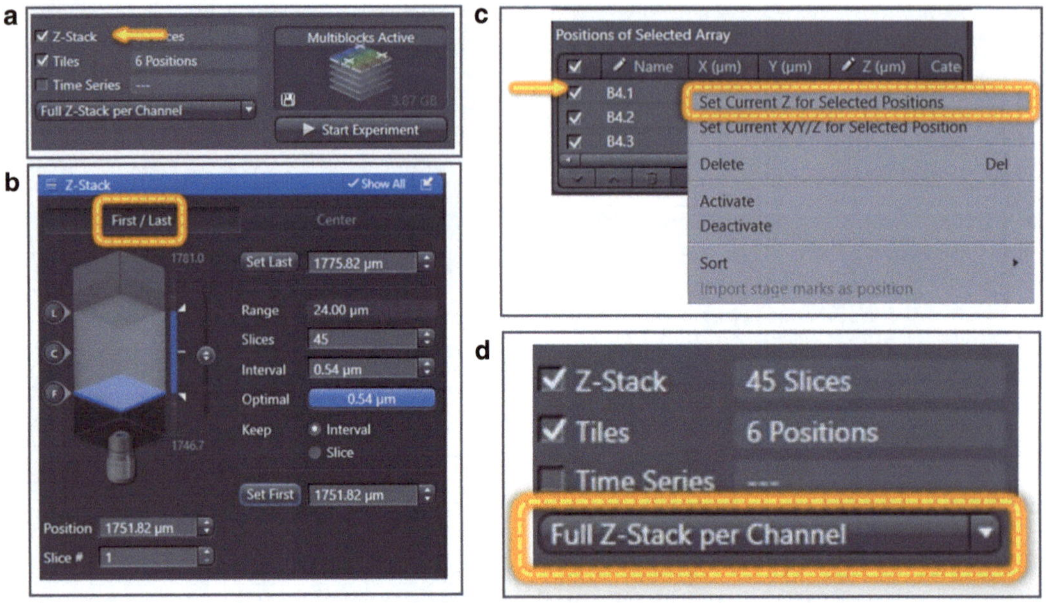

Fig. 4 Setting up the Z-stack. (**a**) Tick the Z-stack button above Tiles. (**b**) Set the Z-stack using the First/Last method and click "C" to center the Z-stack. (**c**) Set the Z-stack for a few positions and note the range. (**d**) Select Full Z-stack per channel to decrease acquisition time

2. Set up the channel intensity based on your sample. Note: Make sure to avoid saturation while still collecting enough signal. As the intensity will be applied across multiple positions on different wells, it is important to verify the positions. In case the intensity varies across different groups (i.e., different combinations of antibodies for staining on the same plate), another acquisition setup can be created for it. See the next section on "Experiment Designer."
3. Set up a Z-stack for the position using the First/Last method, meaning set the first position above the last drusen deposit signal and the last position below the last drusen deposit signal (Fig. 4b).
4. Click on "C" to bring the stage to the center of the Z-stack set up in the previous step.
5. Right-click on the position and select "Set Current Z for Selected Positions" (Fig. 4c).
6. Repeat **steps 3–5** for five to eight positions to get a sense of the range, and take note of the center position.
7. Select the "Center" method on the Z-stack window.
8. Set up the range based on the maximal range found while setting up the different positions above—this is usually around 25–30 µm—and set the optimal interval (e.g., 0.54 µm) for downstream volumetric analysis. The maximal range will ensure that no portion is cut off for image analysis.

9. Toggle to center in the Z-stack tab and set the center position to the average of the sample positions from **step 6**. Set your range based on the highest high and lowest low of your sampled wells. Click C to center the Z-stack (*see* **Note 4**).
10. Under *Channels*, select the ZO-1 channel and along the bottom click *Focus Ref.* Then select *Autofocus*. This will automatically find the center of the well, and take a Z-stack for the designated range/number of slices.
11. Once autofocus is initiated, do a random double check that the cells are in range. If the autofocusing fails, increase the range to allow for detection. Careful consideration needs to be taken for the working distance as a larger range can crash and break the objective. If a well is not, click on it and run autofocus again.
12. Select "Full Z-stack per Channel" for faster Z-stack acquisition (Fig. 4d).
13. Ensure key acquisition parameters are as follows:
    (a) Frame size: 1024 px × 1024 px
    (b) Pixel size: 1.25 μm
    (c) Pixel time: 4.12 μs
    (d) Scan speed: 5
    (e) Direction: bidirectional
    (f) Bits per pixel: 16
14. Click "Start Experiment" to start the acquisition (*see* **Note 5**).
15. Save each Z-stack as a separate file by using Split Scene (Write files) with annotation option in the processing tab; it keeps the Well and Position info in the file name (Prefix-Well-Position.czi).

3.3.5 Setting Up the Experiment Designer (See Note 6)

1. Tick the Experiment Designer tool at the top of the acquisition Table.
2. Experiment Designer window will appear under Multidimensional acquisition (Fig. 5a). Note: Auto Save will be ticked as a default for Experiment designer.
3. Click on "Duplicate" to copy the acquisition settings set in the above sections (Fig. 5b).
4. Select the "2" and modify the positions and intensity based on the experimental group (Fig. 5c).
5. Under the "Auto Save" window, define the folder location and the naming convention. Click on "Start Experiment" to start the acquisition.

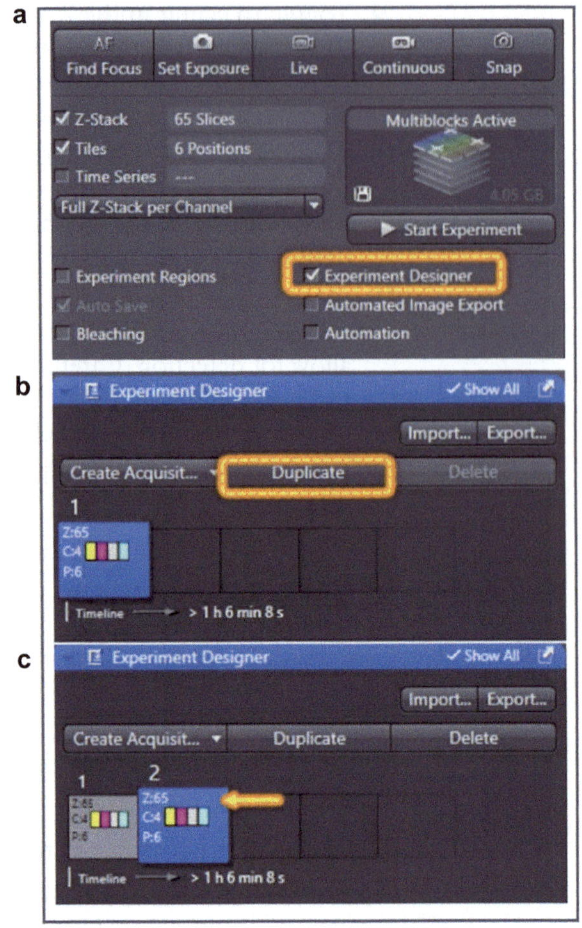

Fig. 5 Setting up multiple experimental blocks. (**a**) Tick the Experiment Designer Box. (**b**) Duplicate the previous block to carry over all settings. (**c**) Click on Block 2 and modify laser intensities as necessary for other antibody combinations or experimental groups

3.4 Imaris Quantification

3.4.1 Development of Batch Parameters

Parameter Determination for Image Processing

1. Open one of the Z-stack images acquired in the LSM900 scan in ImageJ/FIJI.
2. Scroll using the toggles along the bottom of the screen until you find a large deposit that you are confident represents true staining.
3. Use the toolbar along the top of the screen, select *Analyze > Measure*, and draw a line across the deposit (Fig. 6a).
4. A histogram will automatically appear showing the gray values along the line that was drawn (Fig. 6b). At this point it should be obvious what the real signal and the background are.
5. To quantify the background "noise," draw three circles in areas that do not contain any deposits and are thus background (Fig. 6a).

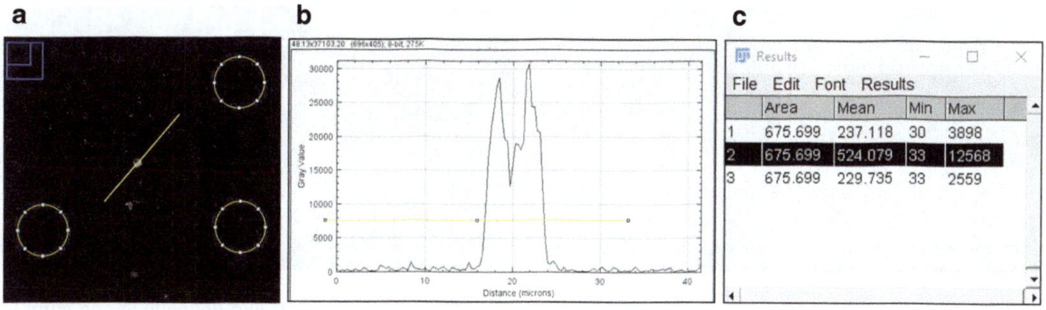

Fig. 6 Background signal elimination quantification. (**a**) Tick the Experiment Designer Box. (**b**) Duplicate the previous block to carry over all settings. (**c**) Click on Block 2 and modify laser intensities as necessary for other antibody combinations or experimental groups

6. A table of results will appear showing the mean, minimum, and maximum values for the signal inside of the drawn circles (Fig. 6c).
7. Take the mean of the maximum values and draw a line along the histogram where that value sits (Fig. 6b). This provides a quantitative basis for background elimination.
8. Repeat **steps 1–7** on the ZO-1 channel and the Hoechst channel.
9. Repeat **steps 1–8** on three to five images until you are satisfied with the gray value that removes background noise but does not eliminate the true protein accumulation, tight junction, or nuclear staining.
10. At the end of this section, you should have a gray value representing the background intensity for APOE (or vitronectin depending on which deposit marker you are working with), ZO-1, and Hoechst.

Determining Batch Image Analysis and Surface Parameters on Imaris

1. Open one of the high-resolution Z-stack Images acquired in the LSM900 scan in Imaris.
2. In the Imaris arena, convert the (.czi) image to an Imaris file format (.ims) using an Imaris File Converter (v9.9.1).
3. Once converted, double-click to open the image in Imaris.
4. *Select File* > *Preferences* > *Statistics* > and tick the boxes next to the following: *Surface* > Center of Homogeneous Mass Z, Position Z, and Volume (3D). Ensure no other boxes are selected as this will make the displayed values much easier to find.
5. Select Image Processing on the top of the screen, and turn off the channels in the *Display Adjustment* window, except for the drusen-like deposit channel (Fig. 7a).

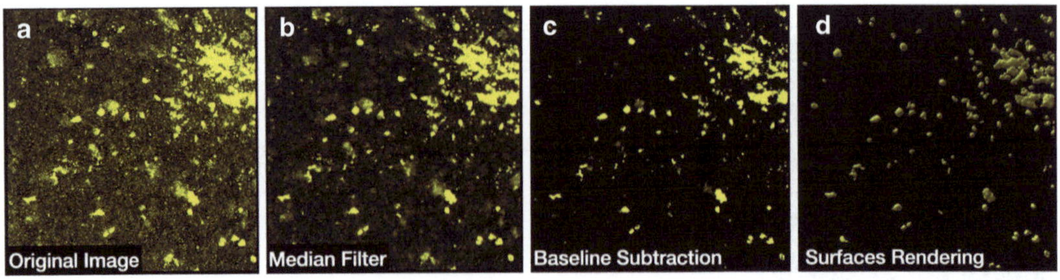

Fig. 7 Image processing for batch analysis from 2D to 3D. APOE stain of drusen-like deposits in one sample. (**a**) Original image prior to any image processing. (**b**) Resulting image after median filter (5 × 5 × 5) is added to the original image. (**c**) Image after the baseline subtraction is applied to the median filter image. (**d**) Final rendering for quantification of drusen-like deposits

6. For each channel (APOE or vitronectin, ZO-1, and Hoechst), apply the following parameters:
   (a) *Median filter*: A median filter can be applied to any channel to reduce background intensity while preserving the edges of the nuclear structure (in our case 3 × 3 × 1 was used for Hoechst and 5 × 5 × 5 was used for APOE/vitronectin and ZO-1). This is a common image processing technique, and there is no need for quantification justification (Fig. 7b).
   (b) *Baseline subtraction*: Set the baseline subtraction for each channel to the gray value that was determined previously (*see* Subheading "Parameter Determination for Image Processing", **step 10**) using a region selection tool to randomly select three background areas to measure the mean intensity in FIJI (Fig. 7c) (*see* **Note 7**).
7. Once baseline subtraction and median filters are added to each of the channels, click okay to process the image at the bottom of the screen.
8. On the top of the screen, click *add Surface (Surfaces*—image segmentation to identify nuclei, cell borders, and drusen-like deposits. This tool is used to render 2D Z-stack Images into a 3D object representing the fluorescent volume in the Z-stack of each channel and used to aid in visualization alongside quantitative analysis including spatial and intensity statistics such as deposit count, cell count, and volume of the deposits).
9. Under Algorithm Settings, tick the boxes next to *Classify Surfaces* and *Object-Object Statistics*. Click the blue arrow to move to the next Table.
10. Under *Segmentation Setup*, select ZO-1, tick the box next to *Smooth*, and leave the default value in *Surfaces Detail* (0.624 in our case), and under *Method* select *Absolute Intensity Thresholding*. Click the blue arrow to move to the next Table.

11. Under *Threshold*, adjust the *Absolute Intensity* histogram until the ZO-1 markers are identified (outlined); in our case the value was 2500. Click the blue next arrow.
12. Under *Filter Surfaces*, adjust the *Number of Voxels* histogram until the ZO-1 marker is defined; in our case the value was 10. Click the blue next arrow.
13. The final tab in the wizard is called *Classification*, and for the ZO-1 channel, you will need to select the trash can to delete Set 1, as nothing in this channel needs to be classified. Click the green next arrow to finish the surface creation. The ZO-1 will now appear as volume-filled 3D renderings.
14. On the top of the screen, click *add Surface again*. Under Algorithm Settings, tick the boxes next to *Classify Surfaces* and *Object-Object Statistics*. Click the blue arrow to move to the next Table.
15. Under *Segmentation Setup*, select the Hoechst channel, select the box next to Smooth, and leave the default value in *Surfaces Detail* (0.624 in our case), and under *Method* select *Absolute Intensity Thresholding*. Click the blue arrow to move to the next Table.
16. Under *Threshold*, adjust the *Absolute Intensity* histogram until every nucleus is identified (outlined); in our case the value was 3500. Click the blue next arrow.
17. Under *Filter Surfaces*, adjust the *Number of Voxels* histogram until the cell nuclei are defined; in our case the value was 444. Click the blue next arrow.
18. The final tab in the wizard is called *Classification*, and for the Hoechst channel, just as the ZO-1 channel, you will need to select the trash can icon to delete Set 1, as nothing in this channel needs to be classified. Click the green next arrow to finish the surface creation. The nuclei will now appear as volume-filled 3D renderings.
19. On the top of the screen, click *add Surface again*. Under Algorithm Settings, tick the boxes next to *Classify Surfaces* and *Object-Object Statistics*. Click the blue arrow to move to the next Table.
20. *Segmentation Setup*, select the drusen deposit channel, select the box next to Smooth, and leave the default value in *Surfaces Detail* (0.624 in our case), and under *Method* select *Absolute Intensity Thresholding*. Click the blue arrow to move to the next Table.
21. Under *Threshold*, adjust the *Absolute Intensity* histogram until every drusen-like deposit is identified (outlined); in our case the value was 3500. Click the blue next arrow.

22. Under *Filter Surfaces*, adjust the *Number of Voxels* histogram until the drusen-like deposits are defined; in our case the value was 444. Click the blue next arrow.
23. Under *Classification* > *Set Name*, select *Position Z* from the drop-down; if this does not appear, ensure you have selected the proper preferences (*see* Subheading "Determining Batch Image Analysis and Surface Parameters on Imaris", **step 4**). Click the green arrow to finish the wizard. The deposits will now appear as volume-filled 3D renderings (Fig. 7d).
24. Repeat all the steps in Subheading "Determining Batch Image Analysis and Surface Parameters on Imaris" for five to eight of your images or until confident that the *Absolute Intensity* and *Number of Voxels* Values *do not eliminate* any real drusen-like deposit, ZO-1 borders, or nuclei.

3.4.2 Setting Up and Running Batch Analysis on Imaris

1. In the Imaris batch arena, click once on any image; this will be the example you will use to set up your batch analysis.
2. In the Imaris arena, click *New Batch*, along the top of the screen.
3. Select *Image Processing* and input the parameters using the optimized image processing and quantification parameters for *baseline subtraction* and *median filter* (*see* Subheading "Parameter Determination for Image Processing", **step 10**).
4. Click add surface and repeat steps (Subheading "Determining Batch Image Analysis and Surface Parameters on Imaris", **steps 8–13**) for ZO-1, Subheading "Determining Batch Image Analysis and Surface Parameters on Imaris", **steps 10–18** for Hoechst, and Subheading "Determining Batch Image Analysis and Surface Parameters on Imaris", **steps 19–23** for drusen-like deposits.
5. Rename each of the Surfaces channels to what they represent (ZO-1, APOE/vitronectin, or Hoechst), and click okay at the bottom of the screen to process the image and create the batch analysis.
6. The batch analysis will now appear in the Imaris arena. Create a new folder and put all images to be analyzed and the batch analysis file (ensure all are in (.ims) file format; if not, convert them before moving to next step).
7. Open the new folder in the batch arena and right-click on the batch analysis file; select Run Batch and the batch processing will begin. The software renders a 3D version of the Z-stack and volume-fills each channel. This allowed for quantification of the amount and volume of the drusen-like deposits (Fig. 8).
8. Export measurements in an excel (.csv) file for further data analysis in PRISM or any other statistics program of choice.

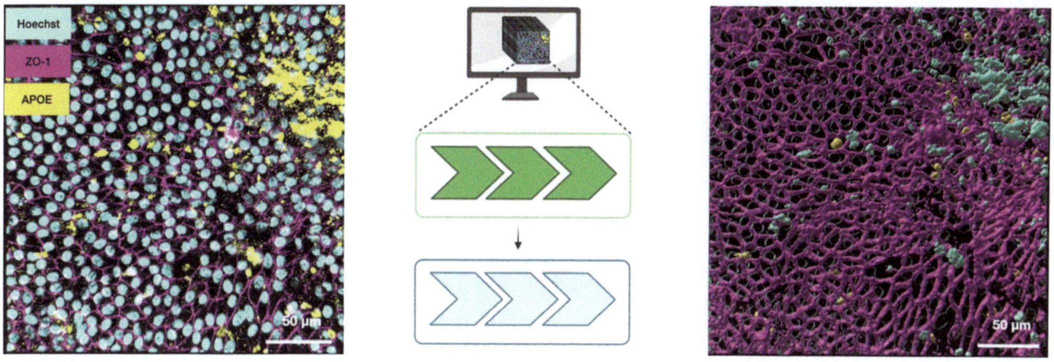

Fig. 8 Batch analysis from 2D to 3D. High-resolution Z-stack with Hoechst, ZO-1, and APOE. The Z-stack is dropped into the batch analysis with optimized parameters, and Imaris generates a three-dimensional rendering of each of the dyes in the Z-stack

This can be done individually or in a batch download from the batch arena.

3.4.3 Setting Up Apical Versus Basal Localization Analysis on Imaris

1. After the Batch Analysis has completed, open one analyzed image (with *Image Processing* and *Surfaces* applied).
2. Click on the *Surfaces* option for ZO-1. Under the statistics tab, find the CHM row under *Variable* and scroll until you see the value for the 75th percentile; record this number.
3. Click on the *Surfaces* option for APOE or vitronectin, select the wand icon and under *Recompute* select *Classification*, and drag the histogram until the line sits at the recorded value for the 75th percentile of the CHM for the ZO-1.
4. Use the ZO-1 and Hoechst Surface channels (ensure the boxes next to them are ticked) to visualize the top of the sample vs the bottom of the sample. The ZO-1 will be above Hoechst, and label your classification under the drusen deposit tab as apical or basal.
5. Export measurements in an excel (.csv) file for further data analysis in PRISM or any other statistics program of choice.
6. This process is manual as the CHM is different for each image; repeat **steps 1–5** for each image.

3.5 Operetta Image Acquisition

1. Open Harmony Software, open the Operetta lid, and load plate into Operetta.
2. In the upper left-hand corner of the screen, fill in the following settings (Fig. 9a):
   (a) Experiment and enter a unique name for your plate.
   (b) Plate type: Select plate type according to the specifications of your plate (*see* **Note 3**).
   (c) Set the objective to 20x high NA.

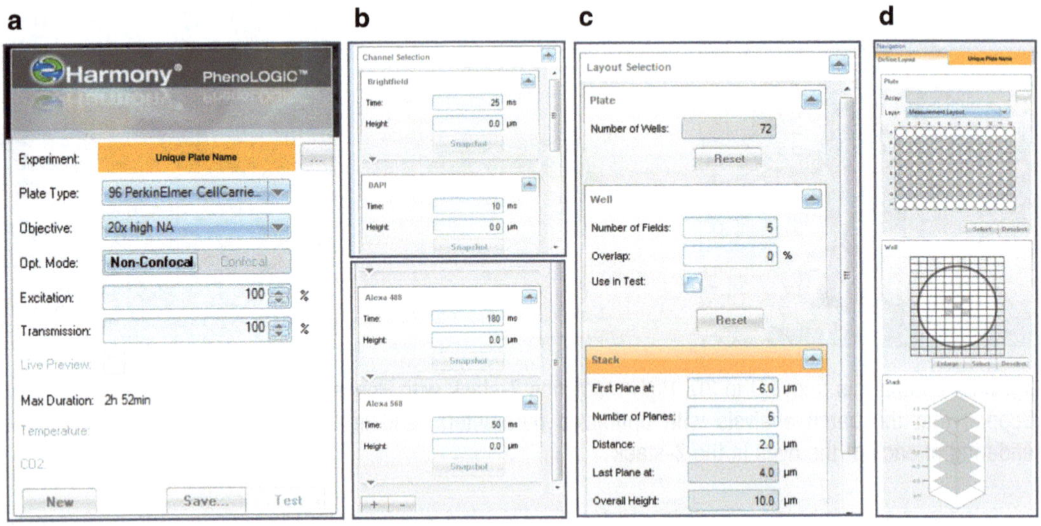

Fig. 9 Setting up Operetta image acquisition parameters in Harmony. (**a**) Harmony Software UI showing user where to enter plate name, select plate type, and select objective. (**b**) Four-channel selection for image acquisition with time and height adjustments provided as examples. (**c**) Layout selection parameters showing five fields of view selected as well as six planes for a Z-stack. (**d**) Final selection parameters under Navigation

3. Add four channels to be collected, in this instance: Brightfield, Hoechst, AF488 (ZO-1), and AF568 (APOE) (Fig. 9b). Parameters are shown in figure but will need to be adjusted to your laser intensity settings (*see* **Note 8**).
4. Fill out Layout selection according to the following parameters (Fig. 9c):
   (a) *Well*
      (i) Number of fields: 5
      (ii) Overlap: 0
   (b) *Stack*
      (i) First plane at: −6 µm.
      (ii) Number of planes: 6.
      (iii) Distance: 2.0 µm.

   This will automatically populate the *Last Plane at* and *Overall Height* categories. Use these parameters as a starting point for your sample, and know the *First plane at* location can be adjusted as necessary for your sample.
5. Under Define Layout, select the wells of interest within your imaging plate; the final setup should appear as is in Fig. 9d.
6. Click Run Experiment tab along the top of the screen. Click start.

3.6 Harmony Quantification

1. Once all wells of your plate have been imaged, click the Image Analysis tab along the top of the screen. The parameters for the Harmony Analysis Sequence are as follows: Basic Parameters (*Input Image: Maximum Projection. Flatfield Correction: None*) (Fig. 10a).
2. The first module finds each cell by using the Hoechst nuclear marker (*Find Nuclei—Channel: Hoechst, ROI: None, Method: C*) (Fig. 10b).
3. The second module uses morphological properties of the detected nuclear that were calculated, allowing removal of artifacts (*Calculate Morphology Properties: Population-Nuclei from previous block, Region: Nucleus, Method: Standard)* (Fig. 10c).
4. Next, any artifact or nonnuclear staining is removed based on size (*Select Population: Nuclei, Filter by property-nucleus morphology* > *0.75*) (Fig. 10d). The output provided here produces a cell count to normalize the drusen-like deposit count to.
5. Next, the software quantifies the deposits (*Find Cells—Channel: AF568, ROI: None, Method: C*) (Fig. 10e).
6. To remove any endogenous intracellular APOE and exclusively count accumulations, the drusen-like deposits are filtered by intensity (*Calculate Intensity Properties—Channel: AF568, Population: Deposits from previous block, Region: Cell, Method: Standard)* (Fig. 10f).
7. The next module filters out any deposits below a certain intensity based on the values from the previous module *(Select Population: Population Deposits, Method Filter by Property: Intensity of the AF568 channel* > usually *600–700* but may need adjusting for your samples) (Fig. 10g).
8. On the final step, the deposit selection is shown, after removal of the low intensity signal, and the user defines the output results, namely, cell count and deposit count (Fig. 10h).
9. On the right-hand *Navigation* panel, click and drag to select all wells, all FOV (5), and all planes (6). Apply image analysis parameters to the entire plate by running "Evaluation." The output is the summation of the maximum projection of five FOV per well.
10. Export the evaluation results as a (.txt) file which can then be imported into an excel file for further data analysis in PRISM or any other statistics program of choice.
11. Either of the two methods can be used for quantification of protein accumulates in vitro (*see* **Note 9**).

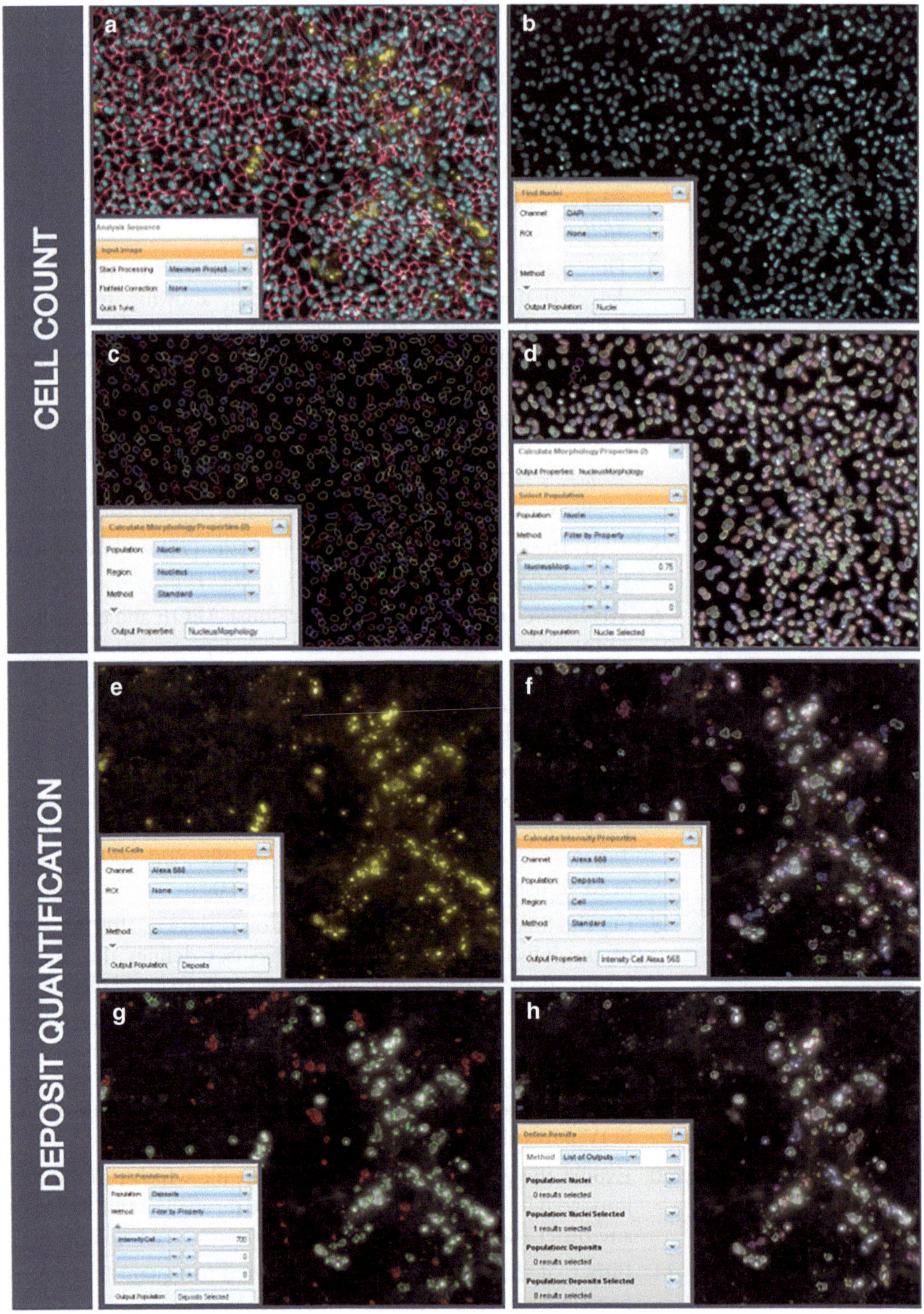

Fig. 10 Cell count and Drusen-like deposit quantification in Harmony. (**a**) Max Projection of iPSC derived with nuclear Hoechst (teal), tight junction marker ZO-1 (pink), and drusen-like deposit marker APOE (yellow), module box showing parameters for Input Image. (**b**) Hoechst-only image, module box showing parameters for Find Nuclei. (**c**) Bounding borders of Hoechst-only image outlined by software to show which nuclei its method

4 Notes

1. Leave one well of each condition to verify the specificity of the staining by the absence of staining in negative controls consisting of the appropriate negative control immunoglobulin fraction in Ms. and Rb. Add the IgG controls to wells at the same time as the primary antibody.
2. Fixed and stained plates can be stored in the final buffer for up to 3 months. However, for best results, aim to image within 5–7 days of staining.
3. Different imaging plates can be customized based on manufacturer specification. In our case, we have customized the dimension for a PhenoPlate 96-well microplates (formerly named CellCarrier Ultra microplates).
4. Using the highest Z position from your tallest sample and the lowest Z position from your shortest sample will create a wide Z-range, but ensures that no apical or basal deposits are missed. You may reduce the Z-range to shorten acquisition time; however, this increases the risk of cutting off deposits and may require reimaging.
5. It is assumed that you have set up your laser intensities for each channel up to this point. Ensure you have made adjustments after assessing three to five wells. It is essential to maintain the same laser intensity values across all wells of samples of the same condition. If laser intensity is changed between samples, the downstream analysis will not work. Keep in mind the best image for analysis is not always the best image for presentations. If some images are overexposed or some images are faint, do not change the laser intensity; once it is set, you must use the same settings for the entire experiment. However, if your plate contains multiple antibody combinations, you can set up an experiment block with different laser settings that can run on the same scan method.

Fig. 10 (continued) detected, module box showing parameters for Calculate Morphology Properties. (**d**) Image of final nuclei selected based on previous two modules and showing module box parameters Select Population which filters out final artifacts. (**e**) APOE-only image channel highlighted, module box showing parameters for Find Cells. (**f**) Bounding borders of APOE-only image outlined by software to show the deposits its method detected, module box showing parameters for Calculate Intensity Properties. (**g**) Bounding borders of APOE-only image outlined by software to show the deposits its method detected, based on intensity selection the green outlined deposits will be counted while the red outlined deposits are removed, module box showing parameters for Select Population. (**h**) Bounding borders of APOE-only image outlined by software to show the final deposits its method detected after intensity threshold is applied, module box showing parameters for Define Results

6. Zeiss LSM 900 confocal microscope was used for this image acquisition, and specifics will be tailored to this microscope although they could broadly be applied to any microscope.

   The most essential elements are as follows:

   (a) Obtain a Z-stack that does not cut off apical or basal deposits; this is usually around 50 μm (~100 sections) Z-stack.

   (b) Ensure the scan speed is slow enough that orthogonal images of the Z-stack are clearly visible as above or below the ZO-1 stain—if acquisition is too fast, they will appear as quite blurry and apical vs basal localization is impossible to determine.

7. An example: The average of three measurements was used as the background value with gray values of 6666 and 1111 measured for the APOE (Alexa Fluor 568) channel and ZO-1 (Alexa Fluor 488), respectively. The Baseline Subtraction tool in Imaris is then used to subtract this baseline value from the intensity of every voxel in the image for the APOE and ZO-1 channels.

8. Select a channel and start with the default settings, and select snapshot below the channel parameters. On the right-hand side, open the histogram for the channel:

   (a) Decrease time if the max intensity is higher than 12,000.

   (b) Increase time if the max intensity is lower than 2000.

   Click snapshot. Repeat until optimal value (histogram mostly sits between 2000 and 12,000). Do this for each channel.

   When to use the LSM900 + Imaris? (Fig. 11)

When quantifying volume or localization of drusen-like deposits

Cons: Very slow due to high resolution necessary; output is per image, not per well or per plate which needs to be calculated separately; and initial setup of batch parameters can take a long time to fine-tune.

Pros: Can distinguish apical vs basal localization as well as provide volume data for drusen-like deposits.

When to use the Operetta + Harmony?

When the only output necessary is the number of drusen-like deposits

Pros: Much faster, covers larger area of the wells, analysis is done on the entire plate, so output of data is already in a plate format so no time lost grouping images per well in analysis, and entire plate is

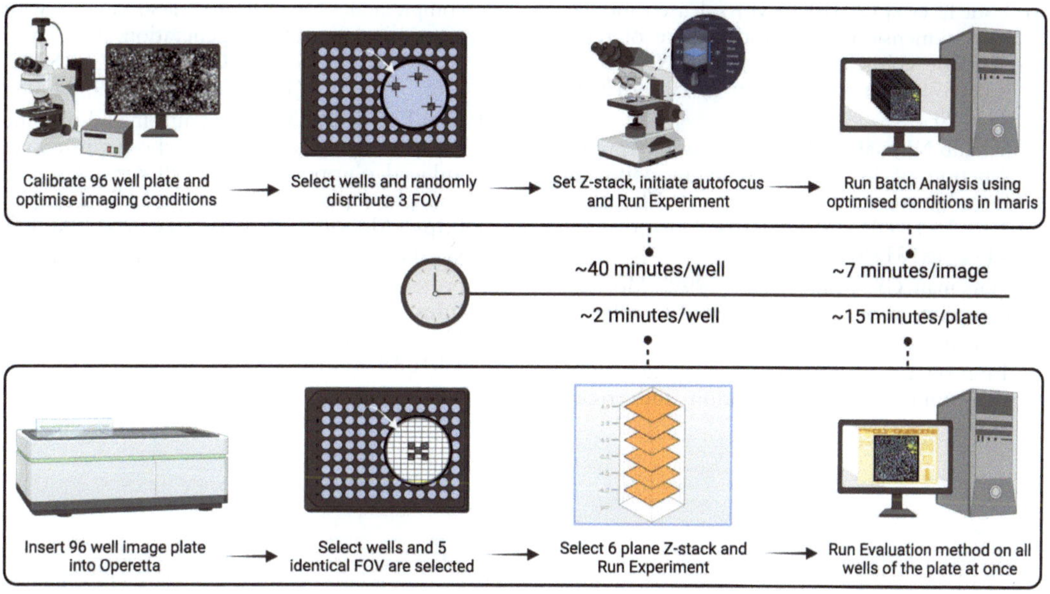

Fig. 11 Comparison acquisition and analysis. Top row LSM900 + Imaris, compared to bottom row Operetta +Harmony

scanned with the same values for lamp intensity so setting up a new experiment is very quick.

Cons: Lose spatial and volume resolution of drusen-like deposits.

Conflicts of Interest

The authors have no conflicts of interest to declare that are relevant to the content of this chapter.

References

1. Jong PTVMD (2018) Elusive drusen and changing terminology of AMD. Eye 32(5): 904–914. https://doi.org/10.1038/eye.2017.298
2. Nathoo NA et al (2014) Optical coherence tomography–based measurement of drusen load predicts development of advanced age-related macular degeneration. Am J Ophthalmol 158(4):757–761.e1. https://doi.org/10.1016/j.ajo.2014.06.021
3. Heesterbeek TJ, Lorés-Motta L, Hoyng CB, Lechanteur YTE, den Hollander AI (2020) Risk factors for progression of age-related macular degeneration. Ophthal Physl Opt 40(2): 140–170. https://doi.org/10.1111/opo.12675
4. Hall JC, Paull D, Pébay A, Lidgerwood GE (2022) Human pluripotent stem cells for the modelling of retinal pigment epithelium homeostasis and disease: a review. Clin Experiment Ophthalmol 50(6):667–677. https://doi.org/10.1111/ceo.14128
5. Galloway CA et al (2017) Drusen in patient-derived hiPSC-RPE models of macular dystrophies. Proc Natl Acad Sci U S A 114(39): E8214–E8223. https://doi.org/10.1073/pnas.1710430114

6. Usui H et al (2018) In vitro drusen model—three-dimensional spheroid culture of retinal pigment epithelial cells. J Cell Sci 132(4): jcs215798. https://doi.org/10.1242/jcs.215798
7. Schaub NJ et al (2019) Deep learning predicts function of live retinal pigment epithelium from quantitative microscopy. J Clin Invest 130(2):1010–1023. https://doi.org/10.1172/jci131187
8. Hageman GS, Mullins RF (1999) Molecular composition of drusen as related to substructural phenotype. Mol Vis 5:28
9. Johnson LV et al (2011) Cell culture model that mimics drusen formation and triggers complement activation associated with age-related macular degeneration. Proc Natl Acad Sci U S A 108(45):18277–18282. https://doi.org/10.1073/pnas.1109703108
10. Hall J et al (2023) A semi-automated pipeline for quantifying drusen-like deposits in human induced pluripotent stem cell-derived retinal pigment epithelium cells. SLAS Technol. https://doi.org/10.1016/j.slast.2023.08.006
11. Senabouth A et al (2022) Transcriptomic and proteomic retinal pigment epithelium signatures of age-related macular degeneration. Nat Commun 13(1):4233. https://doi.org/10.1038/s41467-022-31707-4

Chapter 6

Ex Vivo Culture and Formalin-Fixed Paraffin Embedding of Murine Retinal Explants

Susan Haller, Rebecca M. Marton, and Eliah R. Shamir

Abstract

Traumatic, inherited, and age-related degenerative diseases of the retina are characterized by irreversible loss of retinal neurons, and no present treatments are available to restore them. Cultured murine neuroretinal explants serve as a model of retinal injury and offer a high-throughput, ex vivo approach to test various therapeutic interventions. Here, we describe techniques to isolate and culture retinal explants and treat them with molecules and viruses to enable identification of novel therapeutic strategies. In order to assess the effectiveness of a potential intervention, we also describe protocols to handle friable retinal explants and prepare formalin-fixed paraffin-embedded sections for hematoxylin and eosin staining and immunofluorescence. We also provide tips to overcome common challenges in explant isolation, culture, embedding, and sectioning.

Key words Retina, Retinal explant, Ex vivo culture, Histogel, Formalin-fixed paraffin embedding, Hematoxylin and eosin staining, Immunofluorescence

1 Introduction

The retina is a delicate tissue at the back of the eye that detects photons of light and integrates and transmits this information as electrical signals to the brain. Retinas consist of six major neuronal cell types and one major glial cell type that are produced from a common pool of progenitors that change in competency over time [1, 2]. These neuronal and glial cell types are organized into three nuclear layers and two plexiform layers. Loss of retinal cells or breakdown of this structure can result in permanent visual impairment. One such disease, age-related macular degeneration, involves loss of neurons in the retina and impacts over 150 million individuals worldwide, resulting in a large unmet medical need [3]. While there are presently no therapies available to restore

Susan Haller and Rebecca M. Marton contributed equally to this chapter

Livia S. Carvalho (ed.), *Retinal Gene Therapy: Methods and Protocols*, Methods in Molecular Biology, vol. 3023,
https://doi.org/10.1007/978-1-0716-5198-8_6,
© The Author(s), under exclusive license to Springer Science+Business Media, LLC, part of Springer Nature 2026

neurons after they are lost, advancements in gene therapy have allowed for the correction of photoreceptor dysfunction and the prevention of progressive loss of vision [4]. While promising, presently available gene therapies benefit only a small percentage of patients, and further work is needed to better understand the mechanisms of and identify potential treatments for a wider array of retinal diseases.

One useful tool for studying retinal degeneration and repair is retinal explant cultures [5–7]. In this technique, retinas are dissected from the eye and grown on a membrane in a tissue culture well. Retinal explants have the advantage of allowing direct access to retinal cells while still maintaining the architecture of the retina. With this access, biochemical parameters can be carefully managed, and molecules or viruses can be delivered to retinal cells in controllable concentrations without the risk of impacting other tissues in the organism. This allows investigators to screen potential treatments to inform in vivo study design and minimize adverse effects to living animals. Explant culture also allows for greater observation of cellular dynamics via live-cell imaging [8, 9] and accessibility for electrophysiological measurements of cell function [10, 11].

Following the culture of organotypic explants ex vivo, it can often be useful to examine explants in cross section to assess retinal structure and to perform immunohistochemistry for markers of interest. Processing samples to formalin-fixed paraffin-embedded (FFPE) blocks enables the collection of hundreds of thin (4 μm) sections for downstream assays and provides excellent preservation of morphology. Time in culture and exposure to treatments, however, can make retinal explants friable and difficult to manipulate. The handling and processing steps that are required to create FFPE blocks can further damage delicate explants and lead to tissue loss. These challenges may steer researchers to underutilize FFPE in favor of frozen sections or whole mount immunofluorescence, where the advantages of FFPE are lost. In turn, this protocol builds on our previously published method [12] describing a novel two-step embedding scheme that uses Histogel as a stabilizing agent to improve FFPE explant integrity and therefore section quality. Histogel is a hydroxyethyl agarose product that is commonly used to support embedding of cell blocks or small biopsy samples [13, 14]. Similarly, we use Histogel to offset the friable nature of retinal explants and reduce tissue fragmentation and loss. Our method reliably yields full-thickness retinal cross sections that capture both central and peripheral retinas on one slide [12]. Here, we provide a comprehensive workflow starting from murine retinal explant isolation and culture to embedding, sectioning, hematoxylin and eosin (H&E) staining, and immunofluorescence. These approaches may enable wider use of FFPE in evaluation of retinal regeneration strategies.

2 Materials

2.1 Mouse Retinal Dissection

1. Mice (*see* **Note 1**).
2. Ice bucket.
3. Anlage QTB ready-to-use disinfectant spray.
4. Two Dumont #5 forceps or equivalent.
5. Super fine 8.5 cm, 7 mm tips Vannas spring scissors or equivalent.
6. Dumont #5/45 angled forceps or equivalent.
7. Size 4 paint brush.
8. 1× phosphate-buffered saline (PBS): 800 mL deionized water, 8 g NaCl, 0.2 g KCl, 1.44 g, Na2HPO4, 0.24 g KH2PO4. Do not add Ca^{2+} or Mg^{2+}.
9. 10 cm Petri dish.
10. P1000 pipet and pipet tips.

2.2 Retinal Explant Culture, Treatment, and Sample Fixation

1. Standing cell culture inserts, 30 mm diameter, 0.4 um pore size, transparent polytetrafluoroethylene (PTFE) membrane.
2. Six-well tissue culture-treated cell culture plate.
3. P1000 pipet and pipet tips.
4. P200 pipet and pipet tips.
5. P20 pipet and pipet tips.
6. Retinal explant media: 50% minimum essential medium (MEM)/HEPES, 25% heat-inactivated horse serum, 25% Hanks' balanced salt solution, 200 μM L-glutamine, 5.75 mg/mL glucose, 25 U/mL penicillin, 25 μg/mL streptomycin [15].
7. Phosphate-buffered saline (PBS, without Ca^{2+}, Mg^{2+}). Same as above.
8. 4% paraformaldehyde (PFA): 1× PBS, 40 g paraformaldehyde powder, 1 N NaOH, HCl. Add 800 mL of 1× PBS to a glass beaker on a stir plate in a ventilated hood. Heat while stirring to approximately 60 °C. Take care that the solution does not boil. Add 40 g of paraformaldehyde powder to the heated PBS solution. The powder will not immediately dissolve into solution. Slowly raise the pH by adding 1 N NaOH dropwise from a pipet until the solution clears. Once the paraformaldehyde is dissolved, the solution should be cooled and filtered. Adjust the volume to 1 L with 1× PBS. Recheck the pH and adjust it with small amounts of diluted HCl to approximately 6.9.
9. 70% ethanol.

2.3 Histogel Embedding, Sample Processing, and Paraffin Embedding

1. Histogel (Richard-Allan Scientific, HG-4000-012).
2. Heating block for 10 mL tubes.
3. Incubator oven set at 60 °C.
4. HistoStar Embedding Center (Thermo Scientific) or equivalent.
5. Fine forceps, 5.5″.
6. Artist's brush #1.
7. Biopsy wrap paper, 2″ × 3″.
8. Tissue cassette, 4 cm × 2.8 cm × 0.5 cm.
9. Two beakers: one with 70% ethanol to store tissue cassettes, one for alcohol waste.
10. Laboratory plastic transfer pipets, 0.3 mL.
11. Stainless steel lab spatula, 10″ with rounded end.
12. Dehydrating reagents: 95% and 100%, such as Epredia flex reagent or equivalent (*see* **Note 2**).
13. Xylene.
14. Paraplast regular polymer/wax paraffin, melting point 56 °C.
15. Forceps: 6″, high precision straight broad point or curved medium point.
16. Single edge razor blades, 0.23 mm, with dispenser.
17. Stainless steel embedding molds, 24 × 24 × 5 mm.
18. Black foam tissue sponges, 31.2 × 5 × 2 mm.
19. Disposable absorbent lab cleaning pads.

2.4 Sectioning and H&E Staining

1. Microtome such as HistoCore AUTOCUT microtome (Leica Microsystems) or equivalent.
2. Low-profile, disposable microtome blades.
3. SuperFrost Plus™ microscope slides.
4. 46 °C water bath.
5. Curved forceps, 6.5″.
6. Heated slide plate for drying slides after sectioning.
7. Incubator oven set at 60 °C.
8. Leica AutoStainer XL or equivalent.
9. Xylene.
10. 70% and 95% ethanol.
11. 100% reagent alcohol.
12. Gill's hematoxylin III.

13. 0.185% acid alcohol: 100% ethanol, 37% hydrochloric acid. Mix 950 mL of 100% pure ethanol with 5 mL of 37% hydrochloric acid, and then bring up to 1 L volume with super quality water.
14. Bluing reagent.
15. Eosin Y.
16. Mounting media.
17. Glass coverslips.

2.5 Immunofluorescence

1. 10× concentrate target retrieval solution (Agilent, S169984-2) or equivalent.
2. Deionized water.
3. Slide staining rack.
4. Slide staining jar.
5. DAPI solution: 10 mg/mL 4′,6-diamidino-2-phenylindole (DAPI), water.
6. Tinfoil.
7. Phosphate-buffered saline (PBS, without Ca^{2+}, Mg2+).
8. Normal donkey serum.
9. Triton X-100.
10. Hydrophobic barrier PAP pen.
11. Slide incubation chamber.
12. Bovine serum albumin (BSA).
13. Primary antibodies.
14. Fluorophore-conjugated secondary antibodies.
15. Antifade Mountant.
16. 24 × 55 mm glass coverslip No. 1 thickness.

3 Methods

3.1 Mouse Retinal Dissection

1. Place a membrane insert into each of two wells of a six-well plate with the pronged side down. Add 1 mL of ice-cold PBS to the well below each membrane insert. Keep the six-well plate on ice.
2. Euthanize the mouse in a CO_2-saturated chamber for 5 min followed by cervical dislocation.
3. Clean the dissection area with Anlage QTB.
4. Fill two 10 cm Petri dishes halfway with ice-cold PBS and place on ice in the ice bucket.

5. Remove the left and right eye from the mouse by plucking with angled forceps, and place each eye in separate 10 cm Petri dishes with ice-cold PBS. When removing the eye, clasp the forceps 2–3 mm behind the eye along the optic nerve to retain part of the optic nerve and surrounding tissue with the eye, which aids in dissection.
6. Keep one eye on ice while dissecting the other eye under a dissecting microscope.
7. To isolate the retina, hold the back of the eye/optic nerve with fine #5 forceps (Fig. 1a). With the other hand, puncture the center of the cornea with the tip of one blade of the micro-dissecting scissors (Fig. 1b) (*see* **Note 3**).
8. Cut an opening in the anterior cornea with the spring scissors. The opening should be roughly half of the circumference of the cornea and large enough for the lens and retina to pass through (Fig. 1c, d) (*see* **Note 4**).
9. With one pair of forceps still securely holding the back of the eye, use another pair of fine #5 forceps to gently push in a posterior to anterior direction on the outside of the eye to eject the retina and lens through the opening in the cornea (Fig. 1e).
10. The lens and retina may separate in the process of ejection. If the retina is still attached to the lens, isolate the retina by holding the lens and retina firmly with two sets of forceps and pulling in opposite directions. The suspensory ligaments may tear in the process (Fig. 1f, g) (*see* **Note 5**).
11. Make four equally spaced cuts in the retina with micro-dissecting scissors (Fig. 1h). The cuts should extend one third of the distance from the edge of the retina to the center and should create a clover shape with four petals (Fig. 1i).
12. Trim a P1000 pipet tip to have a ~ 3 mm opening and use the pipet to gently suction up the retina and transfer it to the top of a membrane insert in the six-well plate with the ganglion cell side up (Fig. 1j) (*see* **Note 6**).
13. Arrange the petals to lie flat against the membrane with an artist's paintbrush (Fig. 1k, l) (*see* **Note 7**).
14. Dissect the remaining eye, and place the retina with the ganglion cell side up on a second prepared membrane insert.
15. Repeat **steps 1–14** for additional mice, replacing the PBS in the 10 cm plates with ice-cold PBS between each dissection.
16. When all retinas have been dissected and placed on membrane inserts, replace the 1 mL of PBS below the membrane with 1 mL of retinal explant media pre-warmed to 34 °C, and transfer the plate to a 34 °C incubator with 5% CO_2. Do not exceed 1 h between time of dissection and time of placement in the incubator.

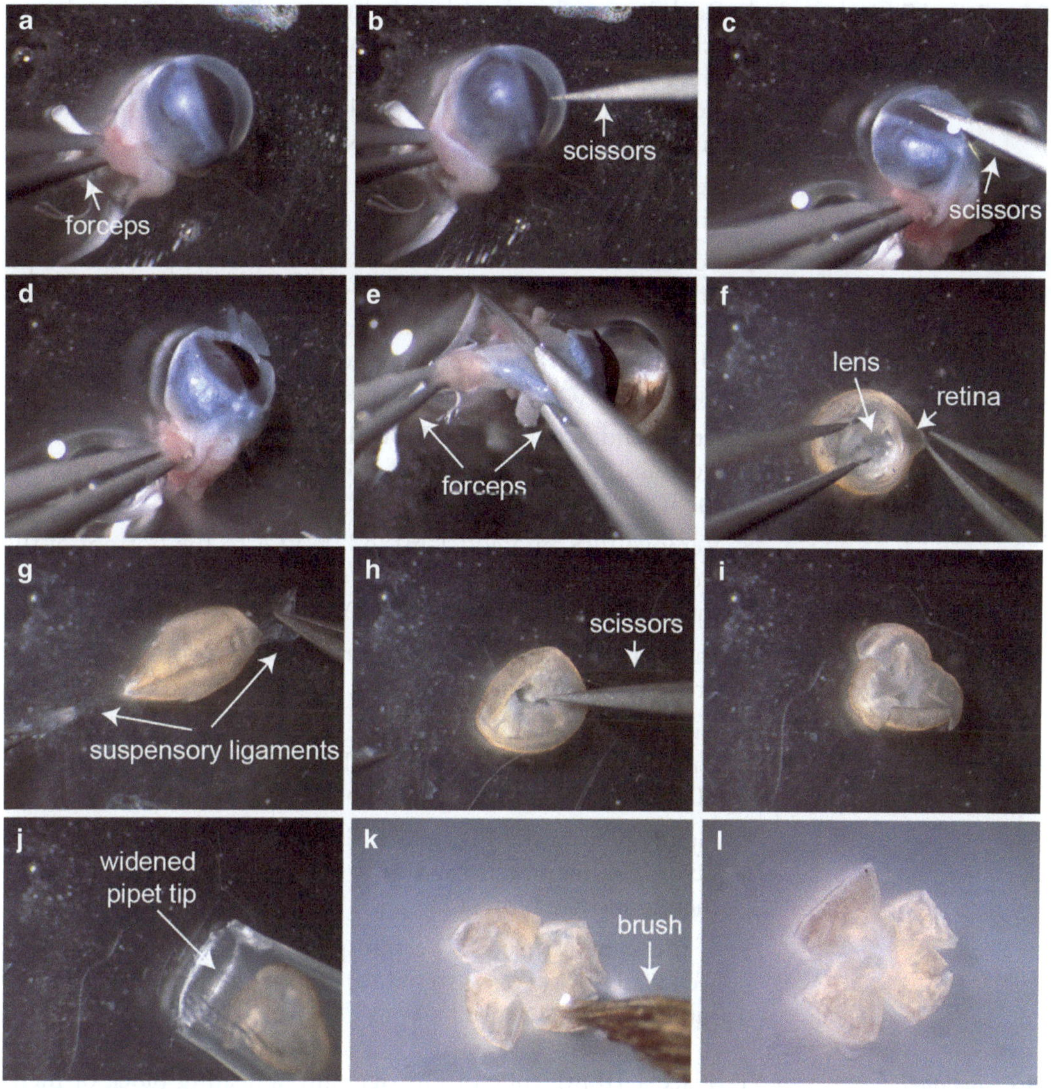

Fig. 1 Mouse retinal dissection. (**a**) The tissue behind the eye is held firmly with forceps to stabilize the eye for dissection. (**b**) The point of a pair of superfine spring scissors is used to puncture the cornea. (**c**, **d**) The scissors are then used to remove roughly two thirds of the cornea to allow the retina and lens to pass through the front of the eye. (**e**) The tissue behind the eye is held firmly with one pair of forceps, while a second pair of forceps gently squeezes the behind the eye and moves in an anterior direction to eject the retina and lens. (**f**) If the lens is still attached to the retina, one pair of forceps is used to stabilize the lens, while a second pair of forceps detaches the retina from the lens by removing the suspensory ligaments. (**g**) Additional suspensory ligaments may need to be removed to allow the retina to be flattened. (**h**, **i**) Scissors are used to make four equally spaced cuts in the retina. (**j**) A widened pipet tip is used to transfer the retina to the culture membrane with the retinal ganglion cell layer up. (**k**, **l**) A paintbrush is used to flatten the petals of the retinal explant on the culture membrane

3.2 Retinal Explant Culture, Treatment, and Sample Fixation

1. Replace the retinal explant culture media every other day by removing the media below the membrane and adding 1 mL of pre-warmed retinal explant media to the bottom of the well. Remove any liquid that has pooled above the membrane without disturbing the explant (*see* **Note 8**).
2. To treat retinas with molecules of interest, add molecules directly to the media at the bottom of the well.
3. To treat retinas with adeno-associated virus (AAV), add 10 μL of virus directly to the top of the retinal explant for ganglion cell-side delivery. For photoreceptor-side delivery, deliver 10 μL of AAV between the explant and the membrane by sliding a P20 pipet tip below the explant (*see* **Note 9**).
4. At the end of culture (*see* **Note 10**), remove all media from the well, and rinse with 1 mL of PBS by adding to the bottom of the well.
5. Remove PBS, and add 1 mL of 4% PFA to the bottom of the well. Add 100 μL of 4% PFA to the top of the membrane over the retinal explant (*see* **Note 11**).
6. Incubate explants in 4% PFA for 20 min at room temperature protected from light.
7. Rinse explants three times for 5 min by removing the 4% PFA and replacing with 1 mL of PBS below the membrane and 100 μL of PBS above the membrane.
8. Gently remove the PBS and replace with 1 mL of 70% ethanol below the membrane and 100 μL of 70% ethanol above the membrane (*see* **Note 12**).
9. If the samples need to be transported, use Parafilm, tape, or well plate sticky covers to reduce evaporation and drying out of samples during transit.

3.3 Histogel Embedding and Sample Processing

1. Place a 10 mL tube of Histogel into a heating block or metal rack in a 60 °C oven for 3 h or until liquified (*see* **Note 13**).
2. Set up the embedding center workstation in preparation for Histogel embedding (Fig. 2): Turn on the embedding center and embedding center cold plate. Prepare the cassettes, biopsy wrap, spatula, a beaker with 70% ethanol, and a beaker for alcohol waste.
3. Take Histogel out of the oven, and leave it in the heating block (*see* **Note 14**).
4. Place the tissue culture plate containing the explants onto the 5 °C cold plate of the embedding center.
5. Use a 0.3 mL transfer pipet to gently aspirate off all of the 70% ethanol from the bottom of the well (Fig. 3a) (*see* **Note 15**).

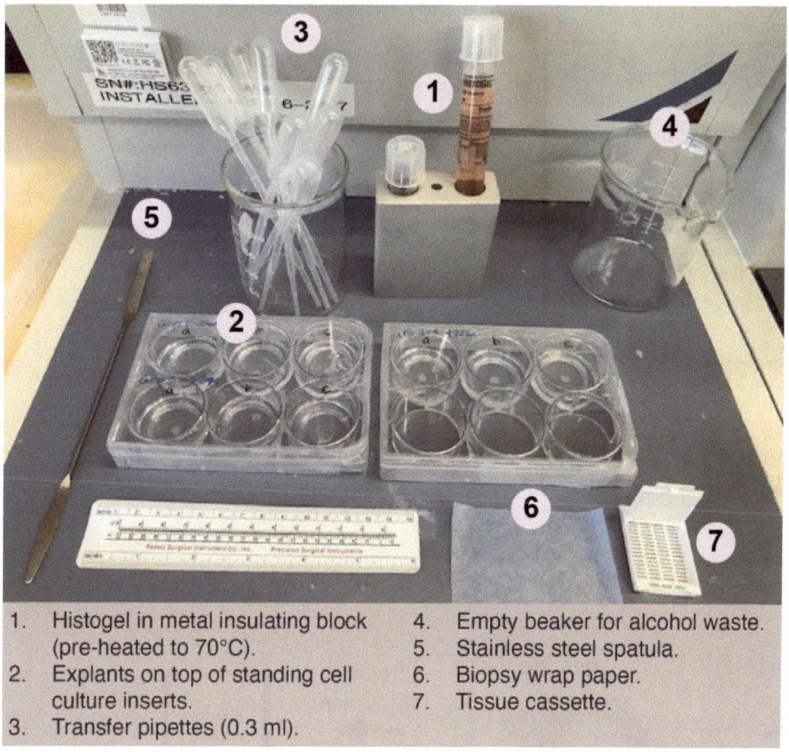

Fig. 2 Workstation set up for Histogel embedding. Layout of all reagents, tools, and equipment used for embedding fixed explants in Histogel

6. Using fine forceps, grip the membrane insert carefully by its edge, lift it out of the well, blot it on an absorbent lab pad to remove excess 70% ethanol, and place it directly onto the 5 °C cold plate of the embedding center (Fig. 3b) (*see* **Note 16**).
7. Center the explant in the middle of the membrane insert using the artist's brush.
8. Slowly add two to three drops (0.1–0.15 mL) of molten Histogel directly on top of the explant using a 0.3 mL transfer pipet (Fig. 3c). Histogel will flow underneath and encase the explant (*see* **Note 17**).
9. After 5 min, use the wider, rounded end of the stainless steel spatula to gently scoop and lift the now firm Histogel-embedded explant onto a biopsy wrap, fold it over letter style, and place it into a tissue cassette (Fig. 3d).
10. Immerse the cassette in a beaker with 70% ethanol, and repeat **steps 6–9** for the remaining samples.
11. Process the cassettes in a tissue processor through graded alcohols, xylenes, and into molten paraffin overnight (*see* Table 1) (*see* **Note 18**).

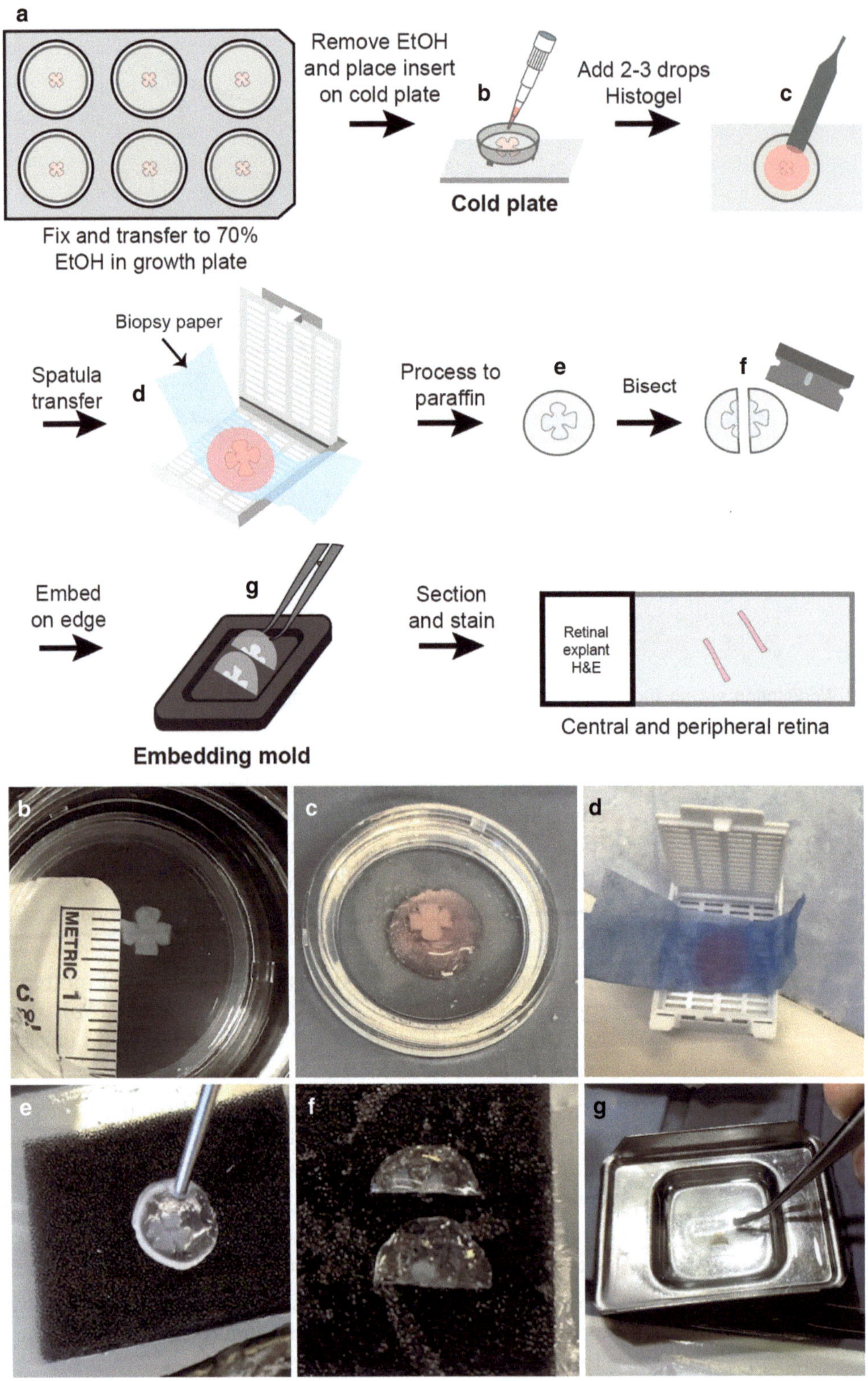

Fig. 3 FFPE processing of murine retinal explants using Histogel. (**a**) Schematic representation of the FFPE processing workflow. (**b**) Retinal explants are fixed on culture inserts in a six-well plate, and the entire insert is transferred onto a cold plate. (**c**) Histogel is pipetted directly onto the explant to encapsulate it. (**d**) A measuring

3.4 Paraffin Embedding

1. Set up the embedding center workstation in preparation for paraffin embedding (Fig. 4a): Turn on the embedding center and embedding center cold plate and cold spot. Place the forceps into a heated forceps holder. Place the sponges and razor blades on the hot surface of the embedding center. Place the embedding molds on the hot surface of the embedding center as well or in the heated tissue storage area.
2. After the tissue has been processed in the automated processor, remove the Histogel-embedded sample from the cassette and place it on the 70 °C hot spot of the embedding center, unwrapping it from the biopsy paper (Fig. 3e) (*see* **Note 19**).
3. Use a razor blade to bisect the explant through the centers of two opposing petals such that each explant half contains both central and peripheral retina (Fig. 3f). Use a fresh razor blade for each sample (*see* **Note 20**).
4. Place the bisected halves on a sponge, carefully gripping each semicircular half at the perimeter using heated forceps to avoid compression of the explant in the center of the Histogel.
5. Dispense approximately 0.5 mL of molten paraffin into a 24 × 24 × 5 mm heated embedding mold, and place it on the 5 °C cold plate of the embedding center to develop a thin film of semi-hardened paraffin.
6. Pick up the first explant half with forceps, and place it diagonally in the embedding mold cut side down (Fig. 3g) (*see* **Note 21**).
7. Once the paraffin begins to anchor the sample to the bottom of the mold, release the forceps and briefly remove the embedding mold from the cold spot (*see* **Note 22**).
8. Pick up the second explant half, place the mold back on the cold spot, and embed the second explant half cut side down, parallel to the first half. Perform this step quickly and fluidly to allow the two explant halves to be embedded in the same plane.
9. After 10 s, lift the mold off of the cold spot, and dispense molten paraffin to fill the mold, placing a labeled cassette lid on top.

Fig. 3 (continued) spatula is used to transfer the embedded explant onto a biopsy wrap, which is folded over and placed into a cassette. (**e**) Following processing, the Histogel is clear and firm. (**f**) A razor blade is used to bisect the explant so that each half contains both central and peripheral retina. (**g**) Each half is embedded on edge diagonally in an embedding mold. (Adapted from Fig. 1 in Haller et al., Improved handling and embedding schemes for cultured murine neuroretinal explants, copyright © 2022 Genentech Inc. © 2022 Journal of Histotechnology, © National Society for Histotechnology, reprinted by permission of Taylor & Francis Ltd., https:/www.tandfonline.com on behalf of National Society for Histotechnology)

Table 1
Tissue processing program[a]

Solution	Time (min)	Temperature (°C)
70% Reagent Alcohol	40	Ambient
95% Flex	40	Ambient
95% Flex	40	Ambient
100% Flex	40	Ambient
100% Flex	40	Ambient
100% Flex	40	Ambient
Xylene	40	Ambient
Xylene	40	Ambient
Xylene	40	Ambient
Paraffin	40	60
Paraffin	40	60
Paraffin	40	60

[a]Vacuum and Agitation Functions OFF

10. Place the mold with the cassette on the large 5 °C plate of the embedding center for 5 min or until the paraffin block is easily lifted out of the mold (*see* **Note 23**).

3.5 Sectioning and H&E Staining

1. Soak FFPE blocks on icy water for 5 min, and section at 4 μm on a rotary microtome using low-profile disposable microtome blades (Fig. 4b, c) (*see* **Note 24**).
2. Pick up the paraffin ribbon using forceps and place sections on a 47 °C water bath for up to 1 min in order to flatten folds in the explant tissue (*see* **Note 25**).
3. Lift sections onto SuperFrost plus slides, air-dry in racks on top of a 60 °C hot plate for 20 min, and then place in an oven at 60 °C for 20 min.
4. For H&E staining, deparaffinize the slides, regressively stain with Gill's hematoxylin III, and counterstain with eosin Y on an automated stainer (Table 2) (*see* **Note 26**).
5. Coverslip H&E-stained slides in mounting media, manually or on an automated coverslipper, and let dry for 1–2 h prior to microscopic examination (representative explant sections shown in Fig. 5).

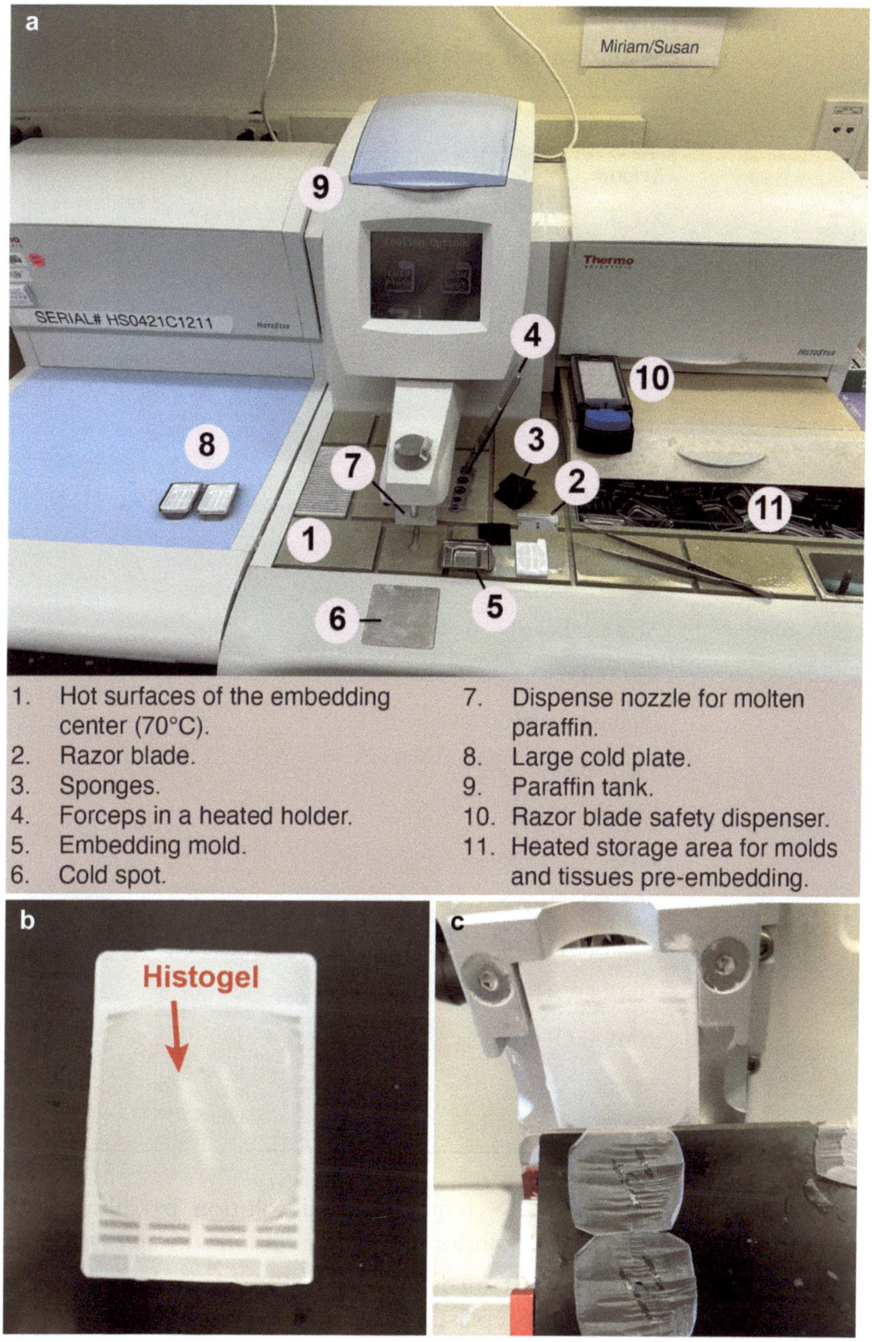

Fig. 4 Workstation setup for paraffin embedding and sectioning. (**a**) Layout of all reagents, tools, and equipment used for embedding processed, Histogel-encased retinal explants in paraffin. (**b**) Representative FFPE block of a retinal explant embedded with Histogel. Histogel appears whiter than paraffin, enabling easy identification of tissue in the block. (**c**) Easy visualization reduces the chance of sectioning past explant tissue. (Adapted from Fig. 2 in Haller et al., Improved handling and embedding schemes for cultured murine neuroretinal explants, copyright © 2022 Genentech Inc. © 2022 Journal of Histotechnology, © National Society for Histotechnology, reprinted by permission of Taylor & Francis Ltd., https:/www.tandfonline.com on behalf of National Society for Histotechnology)

Table 2
Hematoxylin and eosin staining

Solution	Time
Xylene	5 min
Xylene	5 min
100% EtOH	1 min
100% EtOH	1 min
95% EtOH	1 min
95% EtOH	1 min
70% EtOH	1 min
Tap H_2O	1 min
Hematoxylin	8 min
Tap H_2O	1 min
Acid alcohol (0.5%)	30 sec
Tap H_2O	1 min
Bluing reagent	1 min
Tap H_2O	1 min
95% EtOH	1 min
Eosin	30 sec
95% EtOH	1 min
95% EtOH	1 min
100% EtOH	1 min
100% EtOH	1 min
Xylene	1 min
Xylene	1 min

3.6 Immunofluorescence

1. Prepare 1× target retrieval solution by diluting 10× target retrieval solution tenfold in deionized water (*see* **Note 27**).
2. Place unstained slides within a slide rack, and insert the slide rack into a plastic slide jar.
3. Fill the slide jar with 1× target retrieval solution, and place it within a pressure cooker with slide jar adapters. Add several inches of water to the bottom of the pressure cooker.
4. Set the pressure cooker to low pressure mode and 106–110 °C, and allow the slides to process for 10 min after temperature and pressure is reached.

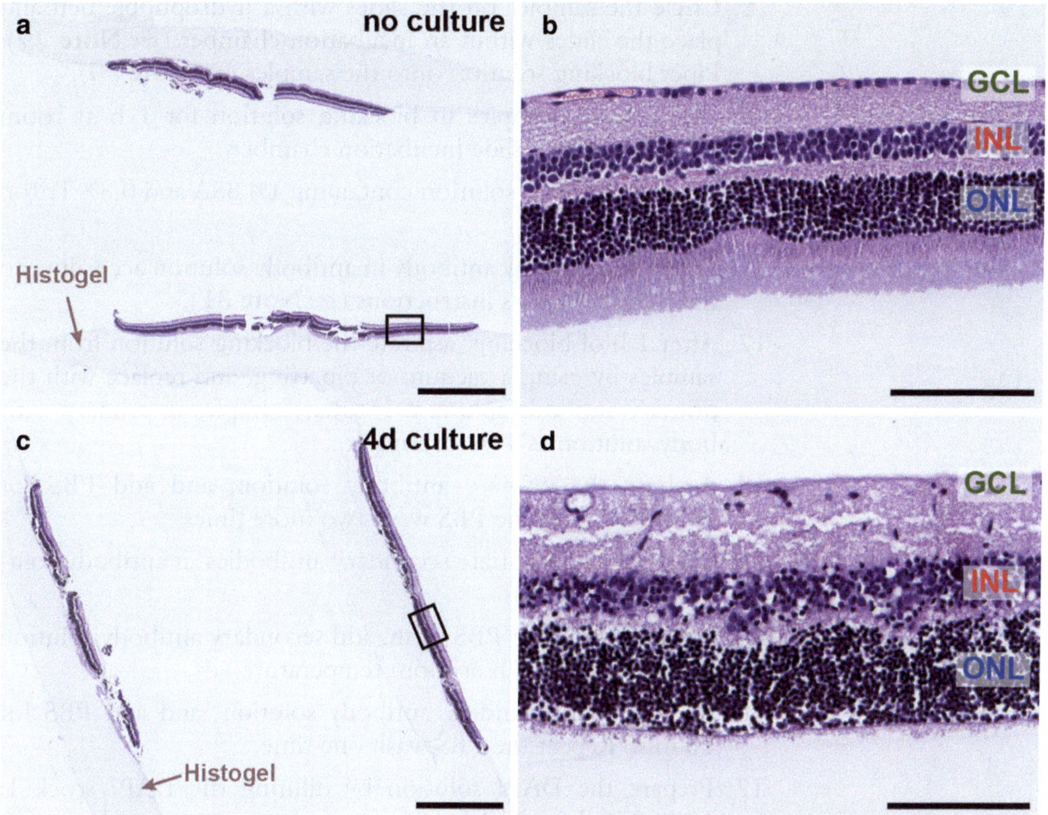

Fig. 5 H&E-stained sections of non-cultured and cultured retinal explants. (**a**, **b**) A representative non-cultured retinal explant showing all layers of the neural retina with normal morphology and cellularity. (**c**, **d**) A representative retinal explant cultured ex vivo for 4 days showing some loss of retinal ganglion cells, degeneration of photoreceptor processes, and cell–cell dyshesion, as is typically seen after several days in culture. Histogel surrounding the explants stains pale pink. Abbreviations: GCL, ganglion cell layer; INL, inner nuclear layer; and ONL, outer nuclear layer. Scale bar = 1000 μm (**a**, **c**); 100 μm (**b**, **d**). (Adapted from Fig. 3 in Haller et al., Improved handling and embedding schemes for cultured murine neuroretinal explants, copyright © 2022 Genentech Inc. © 2022 Journal of Histotechnology, © National Society for Histotechnology, reprinted by permission of Taylor & Francis Ltd., https:/www.tandfonline.com on behalf of National Society for Histotechnology)

5. Remove the plastic staining jar from the pressure cooker, cover the top with tinfoil, and allow the slides to cool at room temperature for 30 min.
6. Rinse the slides by removing the target retrieval solution and filling the jar with PBS. Repeat the rinse two more times (*see* **Note 28**).
7. Prepare blocking solution containing 5% normal donkey serum and 0.5% Triton X-100 in PBS.

8. Circle the samples on the slides with a hydrophobic pen and place the slides within an incubation chamber (*see* **Note 29**). Pipet blocking solution onto the samples (*see* **Note 30**).
9. Incubate the samples in blocking solution for 1 h at room temperature in a slide incubation chamber.
10. Prepare antibody solution containing 1% BSA and 0.3% Triton X-100 in PBS.
11. Dilute the primary antibody in antibody solution according to the manufacturer's instructions (*see* **Note 31**).
12. After 1 h of blocking, aspirate the blocking solution from the samples by using a vacuum or pipetting, and replace with the primary antibody solution. Incubate samples in primary antibody solution at 4 °C overnight.
13. Aspirate the primary antibody solution, and add PBS for 10 min. Repeat the PBS wash two more times.
14. Dilute the appropriate secondary antibodies in antibody solution (*see* **Note 32**).
15. Following the last PBS wash, add secondary antibody solution to the slides for 2 h at room temperature.
16. Remove the secondary antibody solution, and add PBS for 10 min. Repeat the PBS wash one time.
17. Prepare the DAPI solution by diluting the DAPI stock 1: 10,000 in deionized water.
18. Remove PBS and add DAPI solution for 10 min at room temperature.
19. Aspirate the DAPI solution and add one to two drops of mounting media. Apply a coverslip on top of the mounting media, and allow it to cure overnight before imaging. Protect the slides from light.
20. Image slides with an inverted fluorescent confocal microscope (Fig. 6) or other fluorescent microscope of choice (*see* **Note 33**).

4 Notes

1. This protocol has been successfully used to isolate and culture retinal explants from mice ranging from 1 month to 25 months in age. However, protocols were optimized for 6–12-week-old mice.
2. Flex reagents are patented formulations of methyl and isopropyl alcohol that are gentler dehydrants and can minimize over-dehydration and improve sectioning of small, delicate tissues. These are used in place of reagent alcohols.

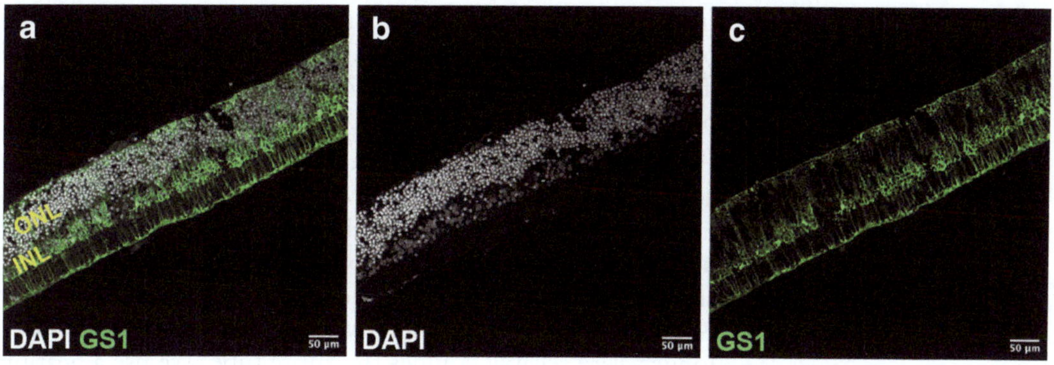

Fig. 6 Immunofluorescence in retinal explant sections. (**a**–**c**) A representative merged image (**a**) of DAPI nuclear stain (**b**) and the Müller glial marker GS1 (**c**) in a retinal explant FFPE section following 7 days of ex vivo culture. Sections were imaged with a Leica inverted SP8 confocal microscope

3. Hypodermic needles or a stab incision scalpel can alternatively be used to puncture the cornea.
4. Take care to avoid cutting the periphery of the retina when cutting the cornea.
5. Further removal of suspensory ligaments may be necessary if the retina maintains a cup-like shape that cannot be flattened (Fig. 1g).
6. If the ganglion cell side is up, the curved tips of the explant petals should be above the retina (curling away from the membrane). If the curled tips of the petals are below the explant, the explant can be flipped over using the paintbrush. Care should be taken not to tear the explant when flipping it.
7. Removal of fluid on top of the membrane can be helpful for flattening the petals of the retina.
8. The media below the membrane may be removed by aspirating from the edges of the well. Lifting the membrane out of the well is not required. If liquid collects above the membrane, this may cause the retinal explant to float off the surface of the membrane.
9. We have only worked with AAVs in Biosafety Level 1 conditions. AAVs carrying hazardous transgenes such as oncogenes, inhibitors of tumor suppressors, or toxins may be classified as Biosafety Level 2 and require additional precautions.
10. We typically culture explants for 4 days. We do not recommend culturing them past 12 days, at which point there is extensive degeneration of retinal ganglion cells, thinning of the inner nuclear layer, and dyshesion within the outer nuclear layer.
11. The 100 μL that is added above the membrane may cause the explant to float slightly off the surface of the membrane. Care should be taken not to jostle the plate if explants are floating.

12. Instructions from the vendor note that samples will embed better in Histogel and be less likely to separate from Histogel if the last solution they are in is 70% ethanol. We typically try to process samples within 2 weeks.
13. Histogel is pink and solid at room temperature. It is stored at 4 °C protected from light until ready for use. We use an oven to liquify the Histogel; other methods suggested by the vendor include microwaving on low power for 5–15 seconds, making sure to loosen up the cap, or placing in a boiling water bath for 3–10 min. We find that the oven method, although it takes longer, very reliably liquifies the entire tube. We use a heating block to hold the tube in the oven and keep the Histogel warm once removed from the oven.
14. If you do not have a heating block for the Histogel tube, it can also be kept warm by storing it in the embedding center's 60 ° C tissue storage area propped inside a large metal embedding mold or metal rack. The cap on the Histogel tube can be removed and left off during embedding for easy pipetting.
15. High volumes of 70% ethanol in the culture plate can cause explants to float off of the tissue culture insert during sample transit and become displaced between the membrane of the insert and the well. If this occurs, rather than lift the explant using an artist's brush, tease the explant flat and embed directly in the well to reduce sample handling. These transport issues are reduced with correct amounts of 70% ethanol and properly sealed plates.
16. To better visualize the explant on the membrane insert, scrape away any icy film on the cold plate to provide better contrast.
17. Care must be taken not to add too much Histogel or to add it too slowly, which can result in variable solidification time and prevent the Histogel from optimally melding together and encasing the sample. This can cause the explant to slip out of the Histogel during or after processing, especially at the bisection step prior to paraffin embedding, often leading to folding artifacts.
18. Turn off vacuum and agitation settings to further protect sample integrity.
19. After processing, Histogel turns from pink to clear, which provides less contrast with the embedded tissue. Moreover, if Histogel starts to cool, it becomes more opaque, and it is even more difficult to see the explant. We have found that the explant is more visible and easier to bisect when the sample is kept in direct contact with the hot plate of an embedding center.
20. The razor blade cuts more cleanly if it is also heated. We rest the razor blade in a groove on the hot area of the embedding

center for approximately 2 min before bisecting the sample. We also find it useful to use a razor blade safety dispenser that requires just a push on the button to dispense a new razor blade. This makes it quicker, safer, and more efficient when bisecting a lot of samples. Alternatively, have razors unwrapped and ready to go before starting, but use caution.

21. To minimize damage to the sample, grip the top of each semicircle, away from the explant, and embed with minimal force.
22. After anchoring the first explant half in paraffin, brief removal from the cold spot ensures that the paraffin does not fully harden before embedding the second half.
23. Embedding two or more tissue samples takes some practice to not create layers in the paraffin due to cooling at different times. The paraffin block should meld together so that, at sectioning, the block does not crumble or fracture at zones where hardening occurred at different times.
24. Histogel appears whiter than paraffin, enabling easy identification of the sample in the block (Fig. 4b). This improved visualization reduces the chance of sectioning past the explant tissue.
25. Experiment with water bath time and temperatures to determine optimal conditions as Histogel tends to separate from the explant at higher temperatures and longer incubation times. Conversely, if the temperature is too low or water bath time too short, folds may be seen in the retina.
26. Histogel appears light pink on H&E-stained sections and in our experience shows no background in immunohistochemistry or immunofluorescence assays.
27. We have found that a deparaffinization step is not needed for our assays of interest, but it could be helpful depending on the target.
28. It can be helpful to vigorously shake the jar from side to side during rinses to knock loose any remaining paraffin on the slides.
29. Drying the slide around the sample can aid in circling the tissue with a hydrophobic pen, but be sure to maintain a thin layer of moisture on the sample itself.
30. The volume of blocking solution applied varies depending on the size of the sample and the area of the circle drawn by the hydrophobic pen. We usually apply 40–50 μL per circled sample.
31. While antibodies vary in affinity, a 1:200 dilution is often a good place to start if no previous data or recommendations from the manufacturer are available.

32. While optimal secondary antibody concentrations may vary between experiments, we usually dilute secondary antibodies 1:1000 in antibody solution.
33. Whole-slide scanning has also been used successfully with this protocol.

Conflicts of Interest

The authors are employees of Genentech, Inc., and are Roche stockholders.

References

1. Cepko CL, Austin CP, Yang X et al (1996) Cell fate determination in the vertebrate retina. Proc Natl Acad Sci 93:589–595. https://doi.org/10.1073/pnas.93.2.589
2. Rodieck RW (1998) The first steps in seeing. Sinauer Associates
3. Wong WL, Su X, Li X et al (2014) Global prevalence of age-related macular degeneration and disease burden projection for 2020 and 2040: a systematic review and meta-analysis. Lancet Glob Health 2:e106–e116. https://doi.org/10.1016/s2214-109x(13)70145-1
4. Gao J, Hussain RM, Weng CY (2020) Voretigene neparvovec in retinal diseases: a review of the current clinical evidence. Clin Ophthalmol (Auckl, NZ) 14:3855–3869. https://doi.org/10.2147/opth.s231804
5. Schaeffer J, Delpech C, Albert F et al (2020) Adult mouse retina explants: from ex vivo to in vivo model of central nervous system injuries. Front Mol Neurosci 13:599948. https://doi.org/10.3389/fnmol.2020.599948
6. Rettinger CL, Wang H-C (2019) Current advancements in the development and characterization of full-thickness adult neuroretina organotypic culture systems. Cells Tissues Organs 206:119–132. https://doi.org/10.1159/000497296
7. Li Y, Zhang Y, Qi S, Su G (2018) Retinal organotypic culture—a candidate for research on retinas. Tissue Cell 51:1–7. https://doi.org/10.1016/j.tice.2018.01.005
8. Lahne M, Li J, Marton RM, Hyde DR (2015) Actin-cytoskeleton- and rock-mediated INM are required for photoreceptor regeneration in the adult zebrafish retina. J Neurosci 35:15612–15634. https://doi.org/10.1523/jneurosci.5005-14.2015
9. Thouvenin O, Boccara C, Fink M et al (2017) Cell motility as contrast agent in retinal explant imaging with full-field optical coherence tomography. Invest Ophthalmol Vis Sci 58:4605. https://doi.org/10.1167/iovs.17-22375
10. Calbiague VM, Vielma AH, Cadiz B et al (2020) Physiological assessment of high glucose neurotoxicity in mouse and rat retinal explants. J Comp Neurol 528:989–1002. https://doi.org/10.1002/cne.24805
11. Alarautalahti V, Ragauskas S, Hakkarainen JJ et al (2019) Viability of mouse retinal explant cultures assessed by preservation of functionality and morphology. Invest Ophthalmol Vis Sci 60:1914–1927. https://doi.org/10.1167/iovs.18-25156
12. Haller S, Marton RM, Marroquin KA, Shamir ER (2022) Improved handling and embedding schemes for cultured murine neuroretinal explants. J Histotechnol 45:182–194. https://doi.org/10.1080/01478885.2022.2119639
13. Shidham VB (2019) CellBlockistry: Chemistry and art of cell-block making—a detailed review of various historical options with recent advances. Cytojournal 16:12. https://doi.org/10.4103/cytojournal.cytojournal_20_19
14. Havnar C, Hotzel K, Espiritu C et al (2022) Standardized processing for formalin-fixed, paraffin-embedded cell pellet immunohistochemistry controls. J Vis Exp. https://doi.org/10.3791/64276
15. Osakada F, Ooto S, Akagi T et al (2007) Wnt signaling promotes regeneration in the retina of adult mammals. J Neurosci 27:4210–4219. https://doi.org/10.1523/jneurosci.4193-06.2007

Chapter 7

Culture of Human Retinal Explants as an Ex Vivo Model for Retinal Gene Therapy

Almaqdad Alsalloum, Madina Khubetsova, Olga Mityaeva, and Pavel Volchkov

Abstract

Optogenetic gene therapy may employ recombinant adeno-associated virus (AAV) vectors for specific preclinical applications in the treatment of inherited retinal diseases. Human retinal explants as an ex vivo model assist determine cellular tropism and the level of subsequent gene expression followed by delivery of viral vectors. Extracts of retinal tissue can be obtained and cultured for a certain time; thus, it is possible to transduce and confirm the effectiveness of the virus using various methods, including immunohistochemical staining. Consequently, we described method for harvesting and culturing human retinal fragments from divergent parts of the retina that differ anatomically and molecularly. The fragments were fixed, sectioned, and stained. A dissociation method has also been described to obtain a retinal cell suspension that can be used in other experiments such as flow cytometry or mRNA/protein level expression assays.

Key words Gene therapy, Human retinal explants, Dissociation, AAV tropism, Culturing

1 Introduction

Optogenetic gene therapy has been presented as a suggestion to remediate the progression of inherited retinal diseases, previously obstinate blindness-related diseases, and even to restore various types of retinal thinning and degradation [1, 2]. Adeno-associated virus (rAAV) vectors for the treatment of retinal diseases have demonstrated safety, efficacy, and an absence of side effects in primary clinical trials conducted by treating patients with *RPE65* gene mutation associated with Leber's congenital amaurosis (LCA2). In addition, improvements in visual field and function have been evaluated in these patients [3, 4].

Animal and cell models have been, and still are, considered the cornerstone for evaluating preclinical AAV-related therapy, despite its lengthy, exorbitant, and indecisive results. The boundless number of AAV designs has made it possible to try out the tropism and

Livia S. Carvalho (ed.), *Retinal Gene Therapy: Methods and Protocols*, Methods in Molecular Biology, vol. 3023,
https://doi.org/10.1007/978-1-0716-5198-8_7,
© The Author(s), under exclusive license to Springer Science+Business Media, LLC, part of Springer Nature 2026

efficacy of AAV in a wide range of animal and cell models; however, there is a discrepancy between the expression of transgenes in the organs of animal models and human subjects [5]. The feasibility of optogenetic gene therapy accrues from the outstanding possessions of the retina, chiefly the unique immune system, the ceiling of the passive prevalence of viral vectors through its own specific barriers, and the feasibility of local injections. On top of that, restoration of visual function following photoreceptor degeneration can be achieved by effortlessly assessing using impartial and subjective functional tests [6, 7]. Jacobson et al. performed authenticated information during LCA2 treatment trials on virus dispensation, toxicity, vector efficacy for targeted cell types, and novel viral vectors in the nonhuman primates. However, nonhuman primate therapeutic evaluation processing is not imperative for the final therapeutic appraisal [8].

Human retinal explants can provide details regarding viral tropism and concerning comparative efficacy [9]. Ex vivo culturing of retinal explants can be obtained in the laboratory with the restoration of functional and morphological features, despite the ineptitude in safety and biodistribution assessments [10, 11]. Human retinal explants are a slightly flawed model depending on the conditions of its extraction, for instance, patients with critical retinal detachment, diabetic eye disease, or vitreoretinopathy [12]. Nevertheless, human retinal tissue is being considered as an eventual preclinical pattern and a target tissue at issue [13]. Johnson and Martin described their own system for culturing rat retinal explants for stem cell therapy [13]. Another study is based on Johnson and Martin system in which human retinal fragments were cultured in a Neurobasal-A based medium after retinal piece extraction at the time of surgery. AAV vectors were constructed and then introduced into retinal fragments with reporter genes to visualize successful transduction of retinal cells using fluorescence microscopy. In addition, post-culturing period experiments including fixation, sectioning for immunofluorescence, or dissociation for processing mRNA/protein expression analysis were obtained [14]. Several basic conditions must be taken into account when extracting and culturing retina, including removal of the anterior part of the eye, the presence of a CO2-independent environment, the careful separation of retinal tissue from the vitreous, and the placement of retinal fragments on a Transwell membrane with a certain pore size [9]. Consequently, we attempted to extract postmortem human retinas from patients' eyes within a specified time frame of 6 hours as maximum to evaluate the tropism and efficacy of viral vectors on retinal cell types. In addition, we obtained a dissociation protocol to prepare cells for analysis of mRNA/protein expression and sectioning of retinal fragments to stain specific retinal cells to determine the affinity of the virus for cell types.

2 Materials

Handling of the human ocular organ should be done in a clean, sterile, and barren environment in a culture hood, which utilizes a positive pressure. Prepared culture media can be pre-aliquoted and stored at −20 °C until use. All surgical and manipulation instruments must be sterilized, especially the retinal fragment dissection tools.

2.1 Dissecting Procedure and Explant Culture

1. Hibernate-A medium (*see* **Note 1**).
2. Retinal culture medium (RCM) contains 500 mL DMEM/F12 (or GMEM) with 5 mL 100× glutamax or L-glutamine, 5 mL 100× nonessential amino acids, 5 mL 100× sodium pyruvate, 4 μl stock β-mercaptoethanol (16 M), 10 mL 50× NS21 supplement, 1.25 mL N-acetylcysteine (0.5 mM), and 5 mL 100× lipid concentrate. Aliquot in 50 mL per tube. Thaw the tube at 4 °C for several hours before use (*see* **Note 2**).
3. Cell culture inserts made of polycarbonate in six-well/12-well plates with a pore size of 0.4 μm.
4. 10 cm diameter Petri dish.
5. 3 mL Pasteur pipettes.
6. Pipettes and filter tips.
7. Humidified incubator set to 37 °C with 5% CO_2.
8. Positive pressure tissue culture hood.
9. Dry ice.
10. Corneoscleral forceps (*see* **Note 3**).
11. Scissors for corneoscleral incision (*see* **Note 4**).
12. Micro-forceps and scissors.
13. Scalpel.
14. 5 or 8 mm diameter round punch biopsy.
15. Surgical spatula.

2.2 Immunofluorescence Tools

1. 4% paraformaldehyde (PFA) in PBS.
2. 30% sucrose solution in PBS.
3. Optimal cutting temperature medium (OCT).
4. Vinyl specimen base molds (15 × 15 × 5 mm).
5. Cryotome.
6. Poly-L-lysine-coated glass slides.
7. Mounting medium.
8. Metal block.
9. Insulated foam container.

10. Liquid nitrogen.
11. Washing buffer: Prepare 0.1% Triton-X, 0.1% Tween-20 in PBS. Use 100 mL PBS as final volume.
12. Blocking buffer: Dissolve in 500 mL distilled water, 50 mL goat serum (S26 liters), 1.25 mL Triton-X, 1.25 mL Tween-20, 11.25 g glycine, 5 g bovine serum albumin (BSA), 5 phosphate-buffered saline (PBS) tablets, and 0.5 g sodium citrate. Make sure the reagents are completely dissolved. Aliquot 5 mL tubes and then freeze at −20 °C for later use (*see* **Note 5**).
13. Staining buffer: Prepare 500 mL distilled water, and then add 1.25 mL Triton-X, 1.25 mL Tween-20, 5 g BSA, and 5 PBS tablets. Aliquot 5 mL per tube and freeze at −20 °C (*see* **Note 5**).

2.3 Dissociation Procedure

1. Retinal ganglion cell (RGC) growth medium.
2. Hanks' balanced salt solution (HBSS) without Ca^{2+} Mg^{2+}.
3. 0.22 μM filter.
4. 5 mL syringe.
5. 40 μm cell strainer.
6. Preparing papain solution: Prepare fresh activation buffer with the following substances and concentrations: 1.1 mM EDTA, 0.3 mM β-mercaptoethanol, and 5.5 mM cysteine-HCl. Use 50 mL HBSS w/o Ca^{2+}, Mg^{2+} and add 110 μL of 0.5 M EDTA, 0.043 g of L-cysteine hydrochloride, and 1 μL of 14.2 M β-mercaptoethanol. Prepare 10 mL of activation buffer, use 1 mg of papain powder, and then filter the papain solution with 0.22 μM filter (syringe/tube). Leave the tube of papain solution open at 37 °C, 5% CO_2 incubator for 30 min before using on the retinal fragment (*see* **Note 6**).

3 Methods

3.1 Retina Explant Collection and Culture

The retina was removed from the donor eye after death (postmortem). For further use, the cornea was cut, and then the retina was isolated. All specimen handling should be performed under aseptic conditions in a positive pressure tissue culture hood.

1. Pour Hibernate-A medium into a 10 cm Petri dish. Place the eyecup in the Petri dish.
2. Remove fat and connective tissue from the eyecup using corneoscleral scissors.
3. Make a circular incision between the transparent cornea and the opaque sclera (through limbus) (Figs. 1a and 2a).

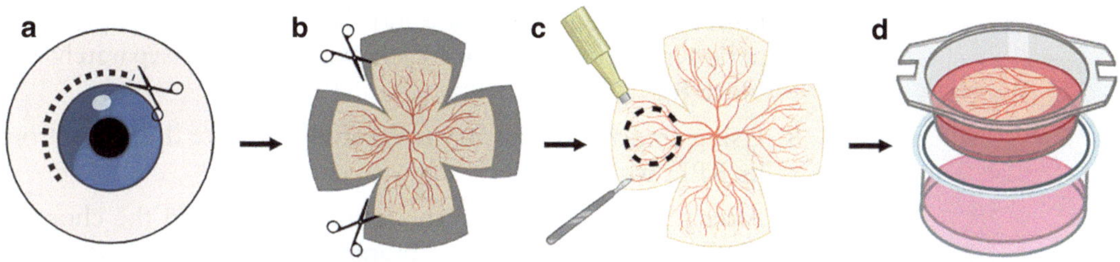

Fig. 1 Simplified scheme for harvesting and culturing retinal explants. (**a**) Obtaining a circular incision between the transparent cornea and opaque sclera and then removing the anterior segment, lens, and vitreous. (**b**) Dividing the remaining eyecup into four center-connected pieces to create a flower-like structure with four petals. (**c**) Diacritic of retinal explants using a 5 mm round punch. Therefore, the explant can be separated from the RPE–choroid structure. (**d**) Transferring of the retinal explant into culture inserts in a well plate followed by culturing retinal tissue for use in gene therapy assessment

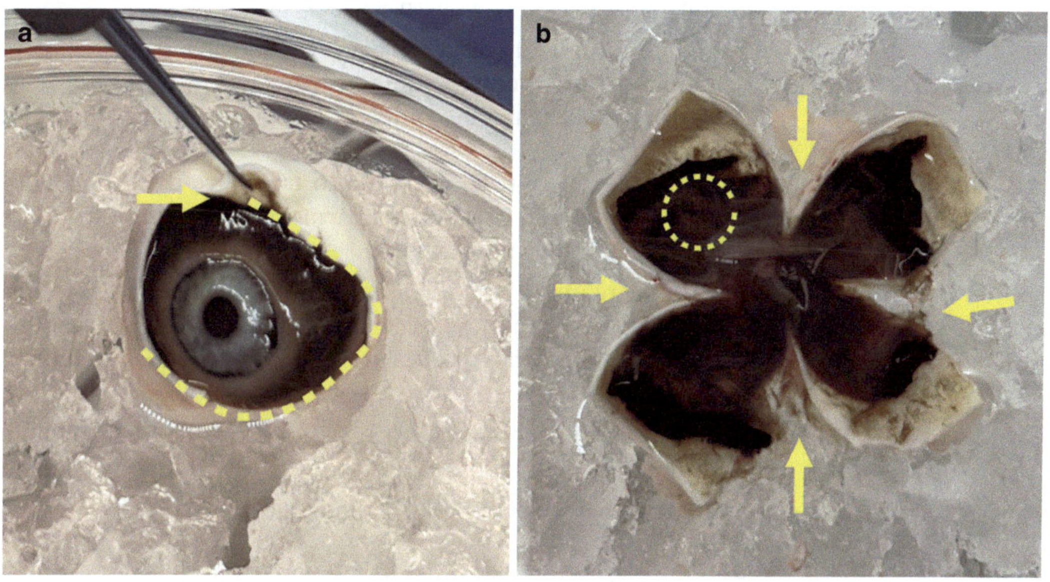

Fig. 2 Retinal dissection and explant culture. (**a**) Human postmortem eyecup after the elimination of fat and connective tissue. A district incision is performed around the limbus of eye (yellow arrow), followed by enucleating the anterior segment and lens. (**b**) The remaining eyecup was quartered (yellow arrows) and flattened to create a flower-like structure with four petals. Therefore, a round punch was used to mark the retinal fragments (dashed line)

4. Remove the anterior segment and lens. Start removing as much as possible from the vitreous body (*see* **Note 7**).
5. Divide the eyecup into four sides to create a flower-like structure with four petals (Figs. 1b and 2b).
6. The flower structure will be clearer when the eyecup is flattened.
7. Use a 5 mm round hole punch biopsy tool for diacritic explants (Figs. 1c and 2b). Press down using the punch and gently roll

until an incision is made around the macula, mid-periphery, or peripheral retina. Culture each type of retina separately (*see* **Note 8**).

8. Using a micro-forceps, try to carefully separate the retina from its RPE, which will be stuck with choroid.
9. After separation, transfer the retinal explant and the choroid-RPE tissue into transmembrane well (Fig. 1d) inserted in six-well plate individually (*see* **Note 9**).
10. Add every three to four pieces of the retinal explants separately per well. Make sure there is enough space between them.
11. Add 1–1.2 mL of retinal culture medium (RCM) to the wells around the transmembrane wells (*see* **Note 10**).
12. Place the six-well plate with the explants in a humidified culture incubator set to 37 °C with 5% CO_2.
13. Change the medium every day by removing the day-before retinal culture medium and add new one. Applying the viral vectors can be performed after 24 h of culturing (*see* **Note 11**).

3.2 Retina Transduction

1. The day after culture, it is time to transduce the retinal fragments with AAV viral vectors. Remove the medium from the well around the culture inserts.
2. Add the required viral particles to the desired part of the retina (macula, mid-periphery, peripheral retina, or RPE–choroid tissue). Use 10^{10} viral particles per 5 mm retinal explant fragment (*see* **Note 12**).
3. After virus injection, add 1–1.2 L of retinal culture medium around the culture inserts.
4. Return the explants back to the incubator at 37 °C with 5% CO_2.
5. Change the medium every day. On the eighth day after transduction, check the fluorescence of the virus (if the viral construction contains a fluorescence reporter gene). Use the explants for the required indications: immunofluorescence or retinal tissue dissociation (*see* **Note 13**).

3.3 Immunofluorescence Staining Procedure

1. After day 8 of transduction (*see* **Note 14**), transfer the retinal or RPE–choroid structure into a tube (*see* **Note 15**), and then wash it with PBS.
2. Carefully aspirate PBS, and then add buffered 4% PFA to the retinal fragments. Incubate overnight at 4 °C (*see* **Note 16**).
3. Remove PFA. Then, wash three times with PBS.
4. Place tissue in 15% sucrose. Incubate overnight at 4 °C (*see* **Note 17**).

5. Change the solution to 30% sucrose (*see* **Note 17**), and store at 4 °C for 1 or 2 days (*see* **Note 18**).
6. Place the metal block into the foam container. Fill with liquid nitrogen so that the block is not completely covered. Wait until the block freezes.
7. Prepare a vinyl specimen mold tray filled with a small-scale amount of OCT.
8. Wash the sample once in PBS. Transfer the retinal fragment to separate vinyl trays (*see* **Note 19**). Cover the top with medium, and do this swiftly so that the sample does not have time to reach the bottom of the tray (*see* **Note 20**).
9. Place the mold tray on the frozen metal block in the container and incubate until the matrix medium is completely frozen (*see* **Note 21**). Trays should be stored at −80 °C prior to sectioning.
10. Mark the edges of the frozen OCT to indicate the location of the specimen.
11. Remove frozen OCT from the plastic mold. Fix the frozen OCT by adding more OCT to the metal platform and along the thin edge.
12. Cut on a cryotome 13 μm thick. Sections should be obtained on glass slides coated with poly-L-lysine (*see* **Note 22**).
13. For immunofluorescence staining, a standard protocol can be followed: Remove the OCT by washing with PBS for 10 min'.
14. Remove excess PBS. Add 200 μL of blocking buffer throughout the slide. Incubate for more than 2 h at ambient temperature.
15. Add the primary antibody at the appropriate dilution to 150 μL of staining buffer.
16. Incubate overnight at 4 °C in a humid plastic chamber.
17. Remove the day-before medium, and wash two times for 10 min with wash buffer.
18. Remove wash buffer, and then add secondary antibody at the appropriate dilution to 150 μL of staining buffer.
19. Incubate for 4 h at ambient temperature in the dark in a humid plastic chamber.
20. Wash as in **step 17**.
21. Add 150 μL of DAPI dye. Incubate for 7 min at room temperature in the dark.
22. Wash the sample for 10 min with PBS.
23. Let the sample dry and cover with mounting medium.
24. Place the coverslip on the glass slide. Allow the mounting medium to cure at room temperature before imaging (Fig. 3).

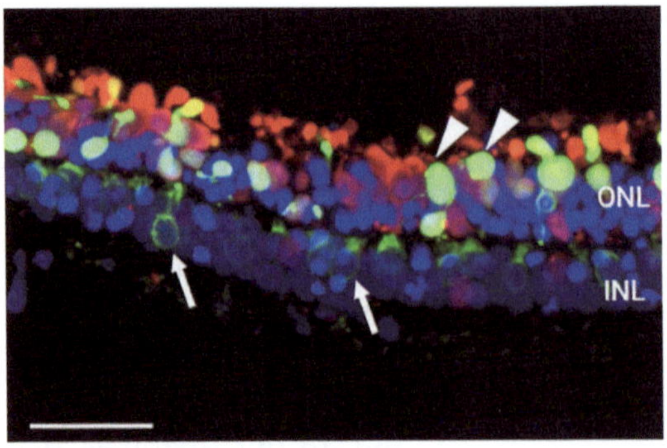

Fig. 3 Transduction of a human retinal explant using the AAV8.GFP viral vector. Confocal microscopy image of retinal explant section after AAV8.GFP transduction, fixation, sectioning, and staining. GFP expression (green) as a marker for transduction of retinal photoreceptors (arrows) and retinal bipolar cells (arrows). Recoverin as a marker of retinal photoreceptors (red). A DAPI fluorescent dye (blue) has been added to indicate the nucleus, scale bar 50 μm. INL inner nuclear layer; ONL outer nuclear layer

3.4 Retinal Explant Dissociation

1. Collect the retinal explant in a 15 mL tube.
2. Settle by gravity for 2 min at approximately 1600–1800 rpm, at ambient temperature to collect retinal tissue.
3. Carefully aspirate the suspension (medium) without losing retinal tissue.
4. Add 8 mL HBSS without Ca^{2+}, Mg^{2+} to the retinal explant precipitate.
5. Centrifuge 7 min at approximately 1600–1800 rpm.
6. Discard the supernatant and resuspend the pellet in papain activation buffer solution (*see* **Note 23**).
7. Incubate retinal tissue with papain activation buffer for 20 minutes in an incubator (37 °C) and shake every 4 min.
8. Gently resuspend the retina with papain activation buffer more than ten times to mechanically disengage the aggregates (*see* **Note 24**).
9. Collect the cell suspension and pass it through a 40 μm cell strainer into a new tube.
10. Centrifuge the sample at approximately 1600–1800 rpm for 5–7 min.
11. Discard the supernatant, and then resuspend the pellet in retinal ganglion cell (RGC) growth medium for culturing (*see* **Note 25**).
12. Calculate the number of retinal cells in suspension and evaluate cell viability.

13. Use the cell suspension as a sample for mRNA/protein expression experiments. Cells can be fixed and stained for other studies such as flow cytometry analysis.

4 Notes

1. It is recommended to use Hibernate-A after the addition of B-27® or 100× L-glutamine; then, it can preserve healthy nerve tissue for up to 1 month at 2–8 °C.
2. For better sterilization and avoidance of contamination, it is recommended to add an antibiotic to the prepared medium.
3. Corneoscleral forceps with a diameter of 0.2 mm tooth (1 × 2 teeth) can be used in sclera dissecting process to successfully hold the sclera during dissection. The teeth can be smaller or larger, and the overall length of the forceps can be about 110 mm for easier handling.
4. One step ahead is the use of corneoscleral scissors with a flat handle and a total length of 110 mm. Scissors may have a curved blade to better cut the cornea around.
5. The frozen aliquot can be thawed before use and stored at 4 °C for up to 2 weeks. It is important to filter the blocking and staining buffer after preparation and before freezing.
6. Measure the PH of the papain solution before and after incubation with the retinal explants. Activation buffer can be prepared and aliquoted and then thawed as needed. The same with papain, it can also be aliquoted and then added to the activation buffer before starting the experiment.
7. In many cases, removing the entire body of the vitreous is quite difficult. The vitreous body, due to its stickiness, can be removed with a Pasteur pipette. Partially aspirate part of the vitreous with a pipette and try to gradually cut off the sucked part until approximately complete removal.
8. Instead of a round hole punch, microscissors can also be used to cut the retina into fragments. Try to gash so that the explants are as similar in size as possible. Folding of retinal fragments can be noticed during the formation of incisions. In these cases, it is recommended not to take pieces for culture. Furthermore, fragments with obvious vitreous impurities are best avoided.
9. Transmembrane wells (cell culture inserts) can be six-well plate or less. Use the appropriate well size according to the experimental conditions.
10. Retinal explants receive nutrition and vital gases from the environment surrounding the pored membrane; therefore, the environment must be sufficient for contact with the

explants through the pores. Avoid high levels of media or bubbles under the bottom surface of the explant, which may affect the viability of the explant and the performance of the experiments.

11. Culture inserts should be moved with sterile forceps when changing media to avoid contamination.
12. Ensure that the added viral particles have spread sufficiently over the surface of the retinal explant. The upper surface is usually photoreceptors, especially when we assume that the results of the assessment will affect the types of photoreceptor cells. An alternative protocol for adding the viral genome is to add the total viral particles at three time points partially prior to day 8 of culture. Days 3, 5, and 7 after transduction are more common in this protocol.
13. Retinal explant size, capsid type, and AAV genome affect transduction efficiency. A negative control is needed to compare the virus signal and can be obtained by culturing explants without viral vectors.
14. Retinal culture can be cultured for up to 2 weeks in many cases. The fluorescence signal stabilized from the eighth day after transduction to the 20th day; in some cases, the cell may begin to gradually degrade after the eighth day.
15. It is important to carefully transfer the explants. Avoid tearing or cruel handling; it is better to use a surgical spatula or a modified Pasteur pipette for transfer.
16. The fixation period and volume should be adjusted for the tissue of interest; for small and thin tissues, 2–4 h at room temperature or overnight fixation up to 4 °C should be sufficient.
17. The retinal explant should be immersed in sucrose to confirm coverage of the entire sample.
18. Avoid contamination while retinal explants are stored in sucrose. It is recommended to start sectioning immediately after a day of treatment with 30% sucrose solution. For longer storage, use 0.1% sodium azide with 30% sucrose and store at 4 °C. Sodium azide is poisonous.
19. Eliminate the PBS around the retinal explant after transferring into the vinyl tray. A Pasteur pipette can be used.
20. The retinal tissue should be oriented. The sample should be laid in the middle of the OCT. Avoid direct tissue contact when repositioning the sample with the pipette tip.
21. Frozen samples can be stored at −80 °C for a longer period of time.

22. The slide can be dried at room temperature during the day and then stored at −20 °C.
23. Use 1 mL papain activation buffer per retina in case of mouse retinal dissociation. It is also recommended to use 1 mL for 5 mm of retinal explant fragment.
24. Carefully resuspend the retina. Avoid excessive agitation which can cause the cells to burst.
25. Alternatively, HBSS can be used to resuspend the dissociated retinal explant pellet.

Conflicts of Interest

The authors have no conflicts of interest to declare that are relevant to the content of this chapter.

References

1. Akimoto M, Miyatake S, Kogishi J, Hangai M, Okazaki K, Takahashi JC, Saiki M, Iwaki M, Honda Y (1999) Adenovirally expressed basic fibroblast growth factor rescues photoreceptor cells in RCS rats. Invest Ophthalmol Vis Sci 40: 273–279
2. Ellouze S, Augustin S, Bouaita A, Bonnet C, Simonutti M, Forster V, Picaud S, Sahel JA, Corral-Debrinski M (2008) Optimized allotopic expression of the human mitochondrial ND4 prevents blindness in a rat model of mitochondrial dysfunction. Am J Hum Genet 83: 373–387
3. Bainbridge JW, Ali RR (2008) Success in sight: the eyes have it! Ocular gene therapy trials for LCA look promising. Gene Ther 15:1191–1192
4. Cideciyan AV, Aleman TS, Boye SL, Schwartz SB, Kaushal S, Roman AJ, Pang J, Sumaroka A, Windsor E, Wilson JM, Flotte TR, Fishman GA, Heon E, Stone EM, Byrne BJ, Jacobson SG, Hauswirth WW (2008) Human gene therapy for RPE65 isomerase deficiency activates the retinoid cycle of vision but with slow rod kinetics. Proc Natl Acad Sci USA 105:15112–15117
5. Asokan A, Schaffer DV, Samulski RJ (2012) The AAV vector toolkit: poised at the clinical crossroads. Mol Ther 20:699–708
6. Gargiulo A, Bonetti C, Montefusco S, Neglia S, Di Vicino U, Marrocco E, Della Corte M, Domenici L, Auricchio A, Surace EM (2009) AAV-mediated tyrosinase gene transfer restores melanogenesis and retinal function in a model of oculo-cutaneous albinism type I (OCAI). Mol Ther 17:1347–1354
7. Komaromy AM, Alexander JJ, Rowlan JS, Garcia MM, Chiodo VA, Kaya A, Tanaka JC, Acland GM, Hauswirth WW, Aguirre GD (2010) Gene therapy rescues cone function in congenital achromatopsia. Hum Mol Genet 19:2581–2589
8. Jacobson SG, Boye SL, Aleman TS, Conlon TJ, Zeiss CJ, Roman AJ, Cideciyan AV, Schwartz SB, Komaromy AM, Doobrajh M, Cheung AY, Sumaroka A, Pearce-Kelling SE, Aguirre GD, Kaushal S, Maguire AM, Flotte TR, Hauswirth WW (2006) Safety in nonhuman primates of ocular AAV2-RPE65, a candidate treatment for blindness in Leber congenital amaurosis. Hum Gene Ther 17:848–858
9. Fradot M, Busskamp V, Forster V, Cronin T, Leveillard T, Bennett J, Sahel J, Roska B, Picaud S (2011) Gene therapy in ophthalmology: validation on cultured retinal cells and explants from postmortem human eyes. Hum Gene Ther 22:587–593
10. Picaud S, Hicks D, Forster V, Sahel J, Dreyfus H (1998) Adult human retinal neurons in culture: physiology of horizontal cells. Invest Ophthalmol Vis Sci 39:2637–2648
11. Pinzo'n-Duarte G, Kohler K, Arango-Gonza'lez B, Guenther E (2000) Cell differentiation, synaptogenesis, and influence of the retinal pigment epithelium in a rat neonatal organotypic retina culture. Vis Res 40:3455–3465

12. Tolmachova T, Tolmachov OE, Barnard AR, de Silva SR, Lipinski DM, Walker NJ, Maclaren RE, Seabra MC (2013) Functional expression of Rab escort protein 1 following AAV2-mediated gene delivery in the retina of choroideremia mice and human cells *ex vivo*. J Mol Med 91:825–837
13. Johnson TV, Martin KR (2008) Development and characterization of an adult retinal explant organotypic tissue culture system as an in vitro intraocular stem cell transplantation model. Invest Ophthalmol Vis Sci 49:3503–3512
14. Orlans OH, Edwards TL, De Silva SR, Patrício MI, MacLaren RE (2018) Human retinal explant culture for *ex vivo* validation of AAV gene therapy. Methods Mol Biol 1715:289–303

Chapter 8

Electrophysiological Characterization of Photoreceptor-like Cells in Retinal Organoids Using Whole-Cell Patch Clamp Method

Almaqdad Alsalloum, Daria Kolotova, Alexey Malyshev, and Pavel Volchkov

Abstract

The patch clamp technique is an electrophysiological method that allows direct measurement of the membrane potential of the cells and current passing through the soma membrane. Photoreceptor-like cells in retinal organoids can be recorded to determine their electrophysiological properties, which are closely related to the physiological functions of mature photoreceptors. Whole-cell patch clamp recording can recognize events of hyperpolarization and depolarization of differentiated cells in retinal organoids. Furthermore, these cells respond to electrical stimulation, as well as to light and darkness. This chapter will explain both types of the whole-cell recordings, voltage clamp and current clamp, and grant the opportunity to characterize the electrophysiological features of presumptive photoreceptor cells. In addition to explaining the patch clamp procedure, the types of solutions used for recording and the manufacture of the recording glass electrodes are described.

Key words Electrophysiology, Patch clamp, Whole-cell recordings, Voltage clamp, Current clamp, Retinal organoids, Photoreceptor-like cell

1 Introduction

The patch clamp method is a pivotal way to assess the cells' response to light by dealing with the changing in the inward/outward ionic currents passing through the cell membrane. The patch clamp makes it possible to assess functional processes in the retina, particularly in the photoreceptors. It also reveals synaptic transport and light responses in photoreceptors to determine their functional maturation [1]. Like other neurons, photoreceptors send signals when changes in their membrane potential occur. Changes in the membrane are driven by the flow of ions stimulated by light, causing hyperpolarization of the photoreceptors. To identify the types of ion channels involved in phototransduction in

Livia S. Carvalho (ed.), *Retinal Gene Therapy: Methods and Protocols*, Methods in Molecular Biology, vol. 3023,
https://doi.org/10.1007/978-1-0716-5198-8_8,
© The Author(s), under exclusive license to Springer Science+Business Media, LLC, part of Springer Nature 2026

photoreceptors, Kawai et al. studied the function of human retinal photoreceptors using the whole-cell patch clamp technique, showing currents such as potassium passing through a pipette into the cell [2].

In response to light, photoreceptors are hyperpolarized, and it was widely believed for a long time that they could not generate action potentials. However, later it has been shown that photoreceptors express voltage-gated sodium channels and can generate action potentials in response to injection of depolarizing current. These action potentials have been shown to be blocked by tetrodotoxin (TTX) and may be involved in the enhancement of the photoreceptor OFF response at the end of the light stimulus during visual perception [2].

Later, Meyer et al. revealed a similarity among the current–voltage relationship that was gained from recoverin-expressing cells in retinal organoids on day 100 of differentiation with what was recorded for mammalian retinal photoreceptors. Subsequently, recoverin-expressing cells were also depolarized by the annexing of a cGMP analog, which was considered an essential influencer on the process of phototransduction. Moreover, they indicate the mimicry of photoreceptor-like cells in retinal organoids with the photoreceptors in regular retina [3, 4].

An administration of 9- or 11-cis retinal into the medium of retinal organoids after 25 weeks of differentiation paved the way for recording the response of photoreceptor-like cells to light. In a different method, the response of photoreceptor-like cells in the 3D structure retinal model displayed a feeble signal compared to the signal obtained from other photoreceptors like human and primates [5, 6]. A specific recording protocol for retinal organoids was obtained by Deng et al. on photoreceptor-like cells, which contain outer segment-like structure after 24 weeks of differentiation. The study analyzed the special electrophysiological features of rod photoreceptors in iPSC-derived retinal organoids by the measurement of hyperpolarization-activated current (I_h) under whole-cell mold using voltage steps that started from −110 mV to 30 mV with a 10 mV accretion [7].

An appointed procedure has been seen in a study of the characteristics of the retinal organoids. It compared the electrophysiological properties of photoreceptor-like cells at three different stages in the maturation pathway of retinal organoids. Approximately on the 180th day of the retinal organoid differentiation, hyperpolarization-activated current (I_h) was measured and compared to the current in primate photoreceptors, showing that I_h increases in parallel with the development of the retinal organoids' capacity and maturation [8]. Consequently, we patched a portion of the photoreceptor-like cells in retinal organoids in order to characterize its electrophysiological activity using the whole-cell patch clamp method. Thus, we obtained a recording of the retinal organoids during its development and maturation into native photoreceptors, determined the currents passing through the membrane, and stained the recorded cells to prove its nature as a photoreceptor-like cell.

2 Materials

The following protocol was optimized for retinal organoids on different stages of maturation. Three main stages can be clearly identified; the early stage is around 80–110 days; the middle stage is 130 days, and the late stage is 150 days and more. Recorded photoreceptor-like cells were located on the external side of the organoids and had a sphere-like, moderately sized shape. During the recordings of photoreceptor-like cells, retinal organoids were perfused in extracellular solution (ACSF, artificial cerebrospinal fluid), at ambient temperature saturated with 95% O2 and 5% CO2. The recording chamber was connected to a superfusion pump system that provided the recorded organoids with ACSF from a reservoir. Experiments were performed with the flow rate of 4–6 mL/min.

2.1 Preparation of the Extracellular Solution

1. An artificial cerebrospinal fluid (ACSF) solution contains 125 mM NaCl, 2.5 mM KCl, 2 mM CaCl2, 1 mM MgCl2, 1.25 mM NaH2PO4, 25 mM NaHCO3, 25 mM D-glucose. The final volume of ACSF is 500 mL. All the solutions are prepared in Milli-Q water. Before obtaining the final volume of ACSF, it is best to make up four types of stock solution. To prepare stock solution I, weigh 92.233 g NaCl, 1.165 g KCl, and 2.126 g KH2PO4. Add water to a volume of 500 mL. Thus, weigh 26.253 g NaHCO3 and add water to 500 mL to make stock solution II. To prepare the $Ca2^{+}$ stock solution, add 7.351 g CaCl2·2H2O to 500 mL volume of water. Weigh 5.08 g MgCl2·6H2O, and then add 500 mL water to prepare Mg^{2+} stock solution (*see* **Note 1**).
2. Take 20 mL of stock solution II into a beaker (*see* **Note 2**), and then add 80 mL water. Bubble the solution with 5% CO2 and 95% O2 for leastwise 15 min.
3. Add 20 mL of stock solution I into a flask, and then add 15 mL volume of Mg^{+2} stock solution. In addition to the previous solutions, add 2.5 g D-glucose.
4. After stock solution II is saturated, transfer stock solution II into the flask with stock solution I, and then add Ca^{+2} stock solution. Make up to 500 mL volume of water (*see* **Note 3**), and then make sure to add a mixture of 5% CO2 and 95% O2 to the final solution during experiments (*see* **Note 4**).

2.2 Preparation of Intracellular Solution

1. The intracellular solution (patch pipette solution) contains 0.3 mM Na2-GTP, 5 mM KCl, 10 mM sodium phosphocreatine, 10 mM HEPES (*see* **Note 5**), 4 mM Mg-ATP, 140 mM potassium gluconate. Using Milli-Q water, prepare a 50–100 mL solution. After all powdered reagents have been added to the solvent, the pH is adjusted to 7.3.
2. Aliquot the solution in a 1.5 mL test tube with a volume of 0.5–1 mL. Store aliquots at −20 degrees.
3. The intracellular solution could be saturated with fluorescent dye Lucifer yellow on the day of the experiment. In this case, fill the pipette with automatic pipette.

2.3 Manufacturing Recording Electrode

1. Choose a pipette made of borosilicate filament glass (*see* **Note 6**).
2. For recording from photoreceptor-like cells, we used capillary glass with an outer diameter of 1.5 mm and an inner diameter of 0.86 mm.
3. Place the chosen glass electrode into a pipette puller. Thus, heat and pull the electrode from the filamented tubing to reach a short gradual narrowing. The average thickness of the electrode should be 10 μm in the middle of its length. Several temperature steps can affect the opening size at the micrometer level (*see* **Note 7**).

2.4 Immunofluorescence Staining of Recorded Cell

1. Paraformaldehyde 4% for retinal organoid fixation.
2. Blocking buffer preparation: A blocking buffer should contain 0.3 M glycine, 0.1% sodium citrate, 0.25% Tween-20, 0.25% Triton-X, 10% goat serum, and 1% bovine serum albumin (BSA) in 1x phosphate-buffered saline (PBS). Take 450 mL of distilled water, and add 50 mL goat serum, 1.25 mL Triton-X, 1.25 mL Tween-20, 0.5 g sodium citrate, 11.25 g glycine, 5 g BSA, and 5 PBS tablets. After a fine dissolving of the reagents, filter the final volume and aliquot it, and then store it at −20 °C (*see* **Note 8**).
3. Staining buffer preparation: A staining buffer contains 0.25% Triton-X, 0.25% Tween-20, and 1% BSA in 1x PBS. Dissolve 1.25 mL Triton-X, 1.25 mL Tween-20, 5 g BSA, and 5 PBS tablets in 500 mL distilled water. After a fine dissolving of the reagents, filter the final volume and aliquot it, and then store it in −20 °C (*see* **Note 8**).
4. Concave microscope slides with one or two round depressions with 0.5–0.8 mm deep.
5. Coverslips about 20 mm wide and less than a millimeter thick.
6. Primary antibodies against specific photoreceptor marker. Secondary antibodies to react with the primary antibodies.

3 Methods

3.1 Preparation of Retinal Organoids

1. Select a retinal organoid at a certain stage of differentiation. We used the organoid at week 22 of differentiation or more for whole-cell patch (*see* **Note 9**).
2. Assess the organoid under visual microscopy to check its condition. The organoid should have a medium-sized symmetrical or uneven spherical shape with a clear formation of the neural retinal layer (Fig. 1b) on the external part of the organoid (*see* **Note 10**).
3. Transfer the retinal organoid to the patch clamp recording chamber filled with extracellular solution. Before fixing the organoids, it is advisable to turn off the perfusion for a short time and only then transfer the organoids into the chamber so that the organoids are not accidentally washed.
4. Fix the selected organoids in the recording chamber using a mesh (Fig. 1a) that can prevent perfusion fluid from moving the organoid in the recording chamber (*see* **Note 11**).
5. Further, only relatively small cells of round morphology and located on the outer surface of the retinal organoids should be selected for the experiment (Fig. 1c). Make sure that the microscope stage does not move until the end of the experiment.

3.2 Whole-Cell Recording Procedure

1. Use a MultiClamp 700B amplifier (Molecular Devices, USA) with pCLAMP 10 software (Molecular Devices, USA). The output should be filtered at 300–10,000 Hz and digitized at 20 kHz (Digidata 1550 Series, Molecular Devices, USA). Turn

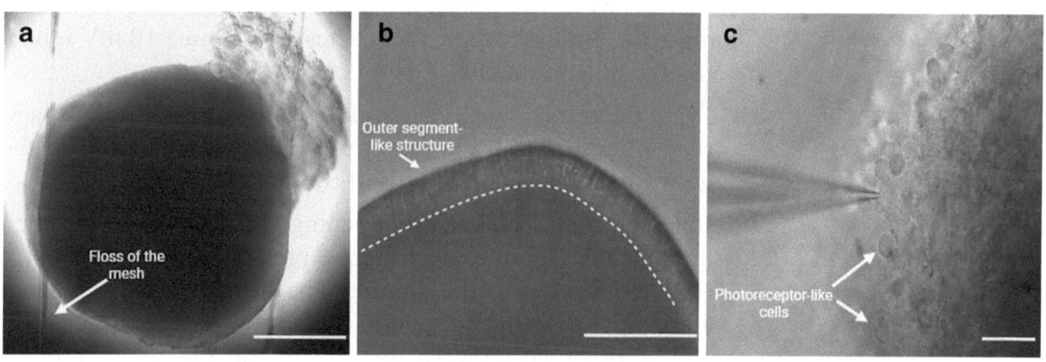

Fig. 1 Bright-field image of a 3D structural retinal organoid at W22. (**a**) Representative image showing the fixation of the retinal organoid with a mesh of thin floss threads. Scale bar, 500 μm. (**b**) Representative images at week 22 showing the thickness of the neural retina (dashed line), which contains an outer segment-like structure on the outer surface of the organoid. Scale bar, 100 μm. (**c**) Bright-field image showing an examination of the electrophysiological properties of OS-containing photoreceptor-like cells in the retinal organoid. Relatively small cells of rounded morphology were selected for registration. Scale bar, 10 μm

on all related equipment, and then turn on the extracellular fluid perfusion pump (ACSF) to perfuse into the recording chamber. The recording chamber should be perfused during the experiment with bubbled extracellular solution (5% CO2, 95% O2) at ambient temperature (*see* **Note 12**).

2. Make a recording pipette using a glass capillary puller Sutter P-1000. The recording pipettes are pulled to a tip resistance of 5–9 MΩ (*see* **Note 13**).
3. Filter the intracellular solution with a 0.2 μm filter. Fill the tip of pipette with 2 μL of pure intracellular solution, and then add 8 μL of intracellular solution saturated with Lucifer yellow dye (*see* **Note 14**).
4. Place the pipette in the electrode holder attached to the head-stage. Ensure that the metal wire inside the recording pipette is in contact with the internal solution.
5. Move the holder forward to orient the electrode and place its tip into the bath in the recording chamber. Avoid poking the microscope lens or the fixed organoid in the chamber.
6. Using a microscope, focus on the tip of the glass capillary and observe its trajectory as it approaches the organoid. Always focus on or below the tip of the glass pipette.
7. Apply slight positive pressure into the pipette. Positive pressure can be obtained through a 1 mL syringe attached to the tube connected to the pipette. Close the three-way valve of the tube to maintain positive pressure (*see* **Note 15**).
8. Point the recording electrode forward into the photoreceptor-like cell, and stop moving the pipette just before approaching the cell. Thus, set the micromanipulator to slow motion.
9. Using the recording software, open the membrane test module function, and then apply the seal test to obtain a 10 mV square voltage pulses at a rate of 100 Hz (*see* **Note 16**).
10. Bring the recording electrode closer to the cell membrane, and then notice a small dimple in the cell membrane due to the light touch between the tip and the cell membrane.
11. Release positive pressure by opening the tube valve, and then immediately apply weak negative pressure by sucking out a tube, connected to the pipette. The tube has an opening to allow the researcher to suction by mouth. As a result, GΩ seal will be obtained between the pipette and the cell membrane (Fig. 2) (*see* **Note 17**).
12. As soon as a GΩ seal has formed, apply strong negative pressure for a very short time in order to rupture the membrane. Set the holding voltage to a potential of −60 mV to −70 mV.

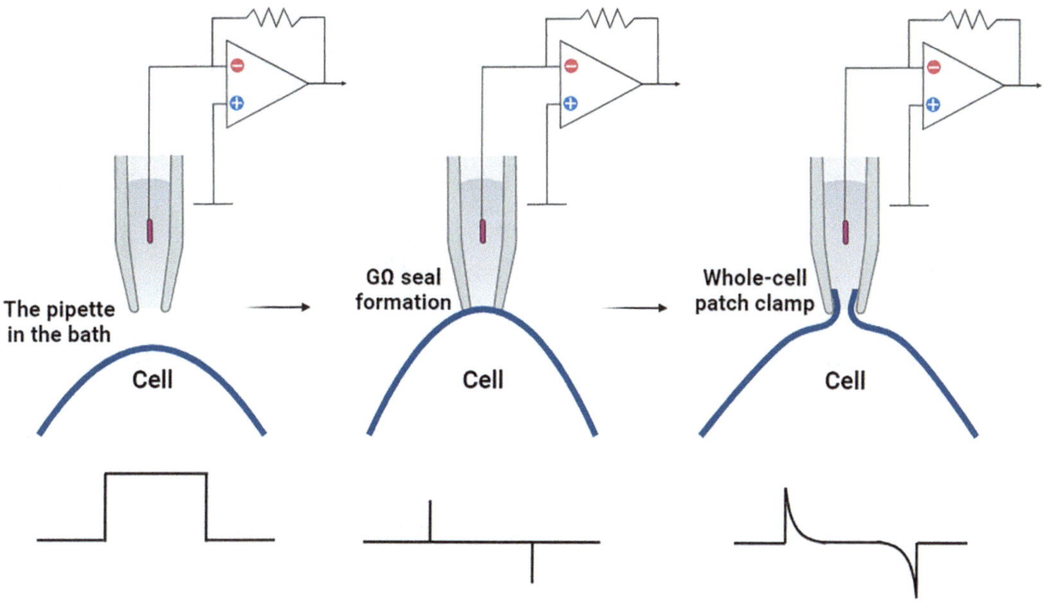

Fig. 2 A simplified scheme of the whole-cell patch clamp procedure. When the pipette enters the bath, the 5–9 MΩ resistance of the pipette results in a nanoampere current. When the cell is attached to the pipette, the pipette capacitance is reset to zero, and the seal resistance in GΩ corresponds to the access resistance. Then the patch section is torn, which will lead to the appearance of larger capacitance. Current characteristic as a result of a membrane resistance of 500 MΩ

3.3 Current Clamp Protocol

1. Switch the amplifier to the current clamp mode.
2. Record the resting membrane potential at zero current.
3. Assess the ability of photoreceptor-like cells to generate spikes by applying multiple positive current steps: 50 pA, 100 pA, 200 pA, and 300 pA of 500 ms duration (Fig. 3) (*see* **Note 18**). Cells often can demonstrate spike-like responses (*see* **Note 19**).

3.4 Voltage Clamp Protocol and Representative Results

1. Switch the amplifier to the voltage clamp mode.
2. Hold the cell at –70 mV.
3. Record whole-cell current responses elicited by voltage steps from –70 mV to +20 mV in 10 mV increments (Fig. 4).
4. Slowly move the pipette tip away from the cell being recorded. Try not to destroy or tear the cell so that you can identify this cell after staining of retinal organoid with antibodies.
5. Repeat the recording of another photoreceptor-like cell in other organoids.

3.5 Staining of Recorded Photoreceptor-like Cell

1. Transfer the retinal organoid from the recording chamber to a tube. Avoid rupture or damage to the organoid (*see* **Note 20**).

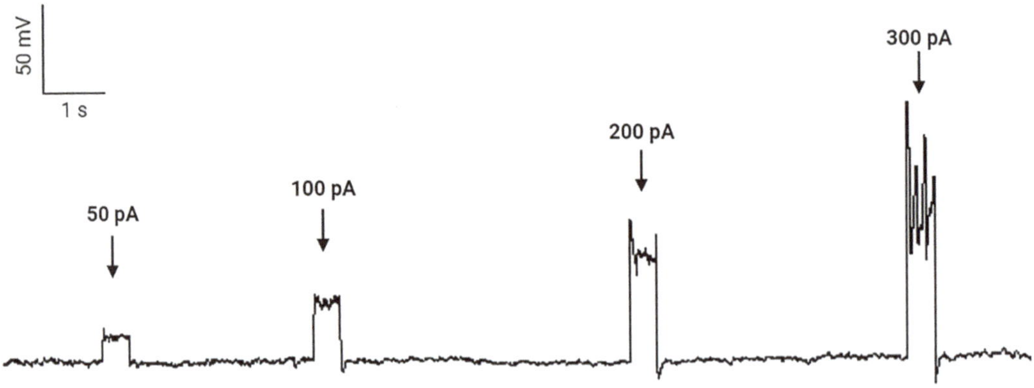

Fig. 3 Investigation of the electrophysiological properties of the photoreceptor-like cell in retinal organoid at week 22 derived from embryonic stem cells. The figure shows the ability of the cell to generate spikes in response to the application of multiple positive current steps: 50 pA, 100 gA, 200 pA, and 300 pA lasting 500 milliseconds. The cell demonstrated a spike-like response when stimulated with 300 pA current

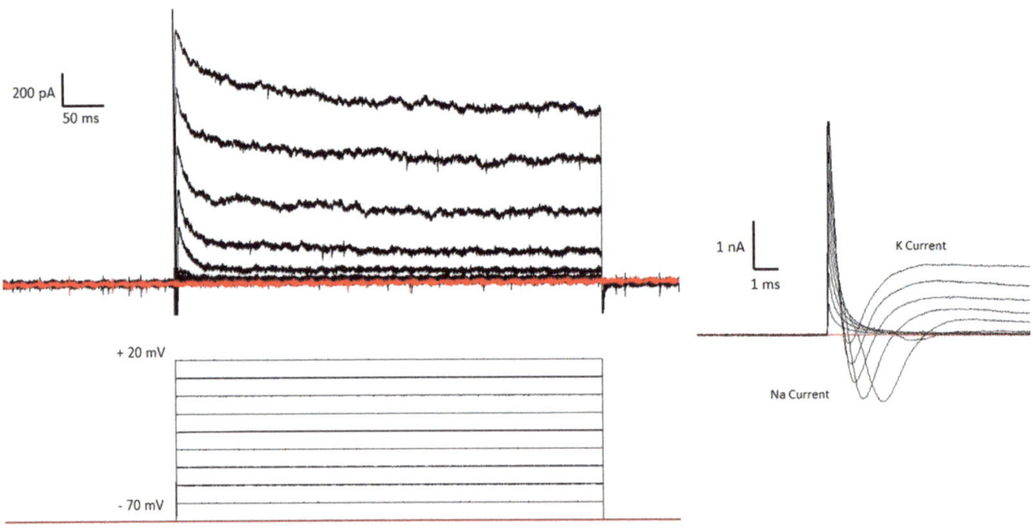

Fig. 4 Average whole-cell patch clamp record was performed in voltage clamp mode from photoreceptor-like cell in retinal organoid at week 22 derived from embryonic stem cells. Representative trace of current responses of photoreceptor-like cell (top left) elicited by voltage steps from −70 mV to +20 mV with a 10 mV increment (bottom left) is shown. Red line highlights the current corresponding to the membrane potential of −70 mV. On the right side, fast currents of sodium and potassium are shown

2. Remove the medium, then add 150–200 μL of 4% PFA, and then incubate the organoid with PFA at room temperature for 20 min.
3. After incubation, wash two times with PBS using an aspirator to remove the surrounding medium.
4. Add 150 μL of PBS to the organoid and store them overnight at +4 °C until the next day.

5. Remove the PBS added the night before. Be careful not to remove any organoid while aspirating.
6. Add 100–150 μL of blocking buffer to the organoid, and then transfer the plate to a shaker for 5 min (*see* **Note 21**).
7. Remove blocking buffer, and then add 100 μL 0.1% Triton-X and 0.1% Tween-20 in PBS as wash buffer. Thus, remove it and wash again.
8. After the second wash, add 100 μL of staining buffer containing the required primary antibodies, such as the anti-recoverin antibody. Incubate the plate overnight at +4 °C.
9. Remove the day-before staining buffer with the primary antibody.
10. Add 150 μL wash buffer for 7 min, and then discard it.
11. Wash again with 150 μL wash buffer for 7 min, discard, and add 100 μL of staining buffer with secondary antibodies (*see* **Note 22**). Incubate overnight at +4 °C.
12. Discard the day-before staining buffer with the secondary antibodies. Add 150 μL wash buffer for 7 min, and then remove it.
13. Add 150 μL wash buffer for 10 min.
14. Wash a third time for 2–3 min, and then remove the wash buffer.
15. Add 100 μL of staining buffer (or PBS) containing DAPI 100x. Incubate for 10 min. Remove the DAPI media.
16. Transfer the organoid to concave microscope slides. Add the mounting medium to the organoid and cover it with a coverslip, and then wait until it dries (Fig. 5).

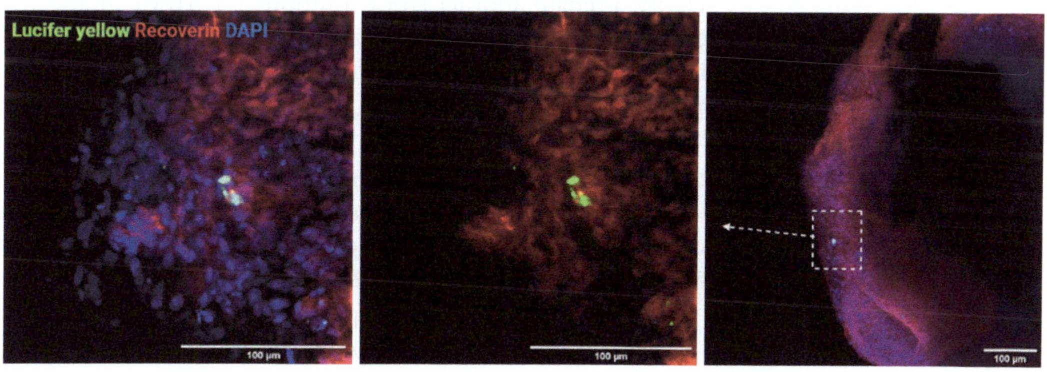

Fig. 5 Representative fluorescence images of a recorded cell located in the outer layer of a 3D retinal organoid derived from embryonic stem cells at 22 weeks. Lucifer yellow was added to the intracellular solution in a pipette before recording the cells; the organoid was fixed and stained with antibodies to recoverin to confirm that the recorded cell was indicative of a photoreceptor-like cell. Scale bar, 100 μm

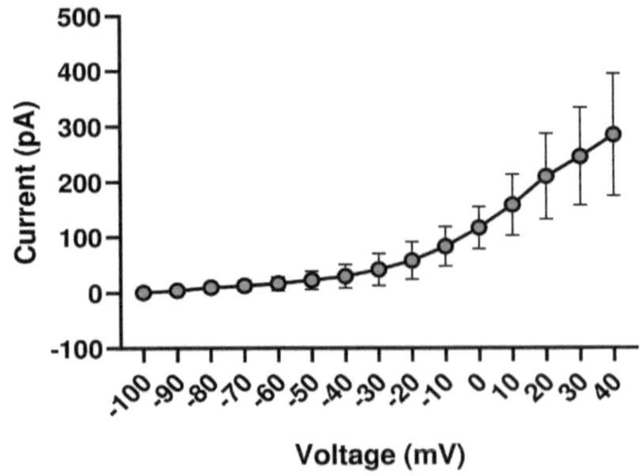

Fig. 6 Average current–voltage (I–V) curve obtained from retinal organoids at week 22 derived from embryonic stem cell. Results are pooled from three independent experiments (*n* = 4)

3.6 Analysis of the Voltage Clamp Recordings

1. Calculate the whole-cell current amplitude for each voltage step for each photoreceptor-like cell.
2. Calculate the average current–voltage relationship (I–V) (Fig. 6) (*see* **Note 23**).
3. Determine other parameters to better characterize photoreceptor-like cells. Check the membrane resistance for each cell depends on Ohm's law. Assess the ease of current flow (conductance) by inverting the resistance of the cell membrane (*see* **Note 24**).

4 Notes

1. A method for preparing ACSF: Make up a 1 L 10X stock solution of NaHCO3, and then store at 4 °C to preserve its activity. Prepare another stock solution: 1 L 10X contains the rest of the reagents, and it should also be stored at 4 °C. From the 10X stocks, prepare 1X solution for the required recording.
2. It is recommended to store the stock solution for up to a several months at 4 °C and prepare a fresh extracellular solution for each individual experiment. While the whole-cell patch may be delayed for several hours, it is best to keep the superfused solution as fresh as possible to obtain the best optimized conditions for organoid cell viability.
3. Ca + 2 stock solution should be added as rapidly as possible into stock solution II. It is advisable to mix the solution with a magnetic stirrer to ensure the homogeneity of the solution.
4. As an alternative for bubbling with CO2 and O2, HEPES-buffered solution can also moderate the PH of the solution.

5. As a retinal organoid is a sensitive cell culture system, it is recommended to use HEPES to stabilize the pH and increase the buffering capacity. HEPES is a zwitterionic N-substituted aminosulfonic acid buffer, thus no need to depend on CO2 levels. HEPES can be made by adding 119.15 g HEPES to 400 mL of water. Then, adjust the pH at ~6.8 by supplying the HEPES with a few pellets of solid NaOH. Adding concentrated NaOH dropwise will amend pH until 7.0. Pour until a final volume of 500 mL is obtained.
6. See the Axon Guide for more profound details on the specifics of glass electrode manufacturing. A wide range of electrode raw materials can be found, but most researchers prefer borosilicate or quartz glass for electrode fabrication.
7. The opening pore at the end of the electrode in a lot of cases is not critical. However, pore size can affect the resistance value; in the case of photoreceptor-like cells, it is important to maintain resistance values below 10–11 MΩ, given that photoreceptors are considered to some extent as small cells. Furthermore, note that the property setting of the pipette puller may vary from device to device, in which the temperature step protocol can directly control the achievement of a fine gradual taper. Before cell recording, it is important to check the electrode tips under a microscope to assess successful heating and pulling.
8. When dissolving, it is better to use a magnetic stirrer. Thawed aliquots of intracellular solution can be stored at +4 °C for up to 1 week; it is recommended to discard the aliquot when it becomes opaque.
9. Determining the exact day to patch the photoreceptor-like cells in the retinal organoids depends on the purpose of the experiment. To characterize the electrophysiological properties of photoreceptor-like cells, it is preferable to choose the day of appearance of recoverin- or rhodopsin-positive cells (*see* Fig. 6). In addition, the formation of cells with an outer segment-like structure is considered as a sign of a primary mature photoreceptor-like cell (*see* Fig. 1b).
10. Good condition of the retinal organoid is important during the patch clamp process, because the cell record is not necessarily obtained promptly; it can take several hours. In addition, the presence of the neural layer of the retinal organoid provides the ability to record a wide range of cell variants (*see* Fig. 1b).
11. The mesh is made of flatted wire in the form of a horseshoe with thin floss threads glued to it.
12. Control the rate of perfusion. In the case of organoid recording, it is better to set the flow rate to 4–6 mL/min to prevent movement of the organoids, mesh, or pipette at a high flow rate.

13. The resistance of the pipette affects the process of achieving and forming a seal. Thinner tips create a high resistance that facilitates seal formation and is better for current clamp protocols but can be problematic for reliable voltage clamp recordings.
14. Filtering the solution is important step to remove excess particles that may interfere with the connection. Make sure that there are no bubbles inside the recording pipette.
15. A small continuous flow of internal solution from the pipette tip can be obtained by positive pressure. All debris and small particles are blown away after the pressure is set. Reduce positive pressure if the cell begins to blow away from the pipette.
16. When we apply the seal test, an amplitude of response appears according to Ohm's law. The amplitude is considered as a response due to the resistance of the pipette: voltage (V) = current (I) × resistance (R). The square voltage current on the oscilloscope is a reflection of the potential between the pipette and the reference electrode after an applied voltage of 10 mV (*see* Fig. 2).
17. The 1 GΩ resistance is a seal property obtained after suction and application of negative pressure. A small current should correspond to the manifestation of GΩ, and the oscilloscope will show an approximately flat response. After sucking out the membrane into the pipette, it is not necessary that the rupture will occur immediately after suction; at least after 1 min, try to make several suctions, one strong one or electrical impulses through the pipette to break in the cell membrane (*see* Fig. 2).
18. Voltage-dependent sodium channel blockers can be added to the extracellular solution during cell recording to assess spike-like events. A common blocker, tetrodotoxin, is used to confirm that the observed spikes are sodium spikes. Furthermore, application of TTX inhibits both I_{Na} and spikes.
19. When we recorded photoreceptor-like cells in the current clamp mode, in most experiments, they demonstrated a non-spike-like response (*see* Fig. 3). However, sometimes neurons respond to intracellular electrical stimulation with time-locked voltage spikes. The researchers also demonstrated differences in the response of different types of retinal cells in their experiments, with retinal pigment epithelium derived from human embryonic stem cells showing no spikes. However, another study showed that bipolar cells respond to extracellular electrical stimulation with voltage spikes. In some cases, the photoreceptors also showed spikes that also require amplitude monitoring in order to understand the response of the photoreceptors to electrical stimulation [9–11].

20. During recording of the photoreceptor-like cell, the dye Lucifer yellow in the intracellular solution will penetrate through the tip into the cell. Consequently, Lucifer yellow will work as an indicator to the presumptive photoreceptor-like cell (*see* Fig. 6).
21. It is best to warm the blocking buffer up for at least 1 h at room temperature before adding.
22. Because Lucifer yellow excites with 488 nm wavelength lasers, select the second antibody's fluorophore to excite with a different laser wavelength.
23. Current–voltage (I–V) relationship is considered as a pivotal parameter for assessing the electrophysiological properties of photoreceptor-like cells due to the ability to visually determine the direction of the current flow at a certain membrane potential. The use of electrophysiological voltage clamp techniques allows the I–V curves to be set to match the desired membrane voltage, so that the appropriate current can be recorded after experimental voltage adjustment. However, coverage of the physiological voltage range should be experimentally confirmed with a typical voltage range for the I–V curve from −150 mV to +50 mV.
24. In a cell membrane, conductance is the pathways that allow ions to pass through the membrane. This refers to the measurement of the ease of ion passage through the membrane. Conductance is the inverting value of resistance, so it is related to current and voltage: $G = 1/R$, $I = GV$, where I is current (in amperes), G is conductivity (in siemens), V is voltage (in volts), and G is resistance (in ohms).

Conflicts of Interest

The authors have no conflicts of interest to declare that are relevant to the content of this chapter.

References

1. Kornreich BG (2007) The patch clamp technique: principles and technical considerations. J Vet Cardiol 9:25–37
2. Kawai F, Horiguchi M, Suzuki H, Miyachi E (2001) Na^+ action potentials in human photoreceptors. Neuron 30:451–458
3. Neher E, Sakmann B (1976) Single-channel currents recorded from membrane of denervated frog muscle fibres. Nature 260:799–802
4. Meyer JS, Howden SE, Wallace KA, Verhoeven AD, Wright LS, Capowski EE, Pinilla I, Martin JM, Tian S, Stewart R, Pattnaik B, Thomson JA, Gamm DM (2011) Optic vesicle-like structures derived from human pluripotent stem cells facilitate a customized approach to retinal disease treatment. Stem Cells 29:1206–1218
5. Zhong X, Gutierrez C, Xue T, Hampton C, Vergara MN, Cao LH, Peters A, Park TS,

Zambidis ET, Meyer JS, Gamm DM, Yau KW, Canto-Soler MV (2014) Generation of three-dimensional retinal tissue with functional photoreceptors from human iPSCs. Nat Commun 5:4047

6. Baylor DA, Nunn BJ, Schnapf JL (1984) The photocurrent, noise and spectral sensitivity of rods of the monkey Macaca fascicularis. J Physiol 357:575–607
7. Deng WL, Gao ML, Lei XL, Lv JN, Zhao H, He KW, Xia XX, Li LY, Chen YC, Li YP, Pan D, Xue T, Jin ZB (2018) Gene correction reverses ciliopathy and photoreceptor loss in iPSC-derived retinal organoids from retinitis pigmentosa patients. Stem Cell Rep 10:1267–1281
8. Li L, Zhao H, Xie H, Akhtar T, Yao Y, Cai Y, Dong K, Gu Y, Bao J, Chen J, Zhang M, Zhong K, Xu W, Xue T (2021) Electrophysiological characterization of photoreceptor-like cells in human inducible pluripotent stem cell-derived retinal organoids during in vitro maturation. Stem Cells 39:959–974
9. Belmonte C, Stensaas LJ (1975) Repetitive spikes in photoreceptor axons of the scorpion eye. Invertebrate eye structure and tetrodotoxin. J Gen Physiol 66:649–655
10. Ignatova I, Frolov R, Nymark S (2023) The retinal pigment epithelium displays electrical excitability and lateral signal spreading. BMC Biol 21:84
11. Walston T, Chow R, Weiland J (2015) Patch clamp recordings of retinal bipolar cells in response to extracellular electrical stimulation in Wholemount mouse retina. Conf Proc IEEE Eng Med Biol Soc 2015:3363–3366

Chapter 9

Generation and Characterization of Patient-Derived Human Pluripotent Stem Cell Lines for Retinal Gene Therapy

Nathan Ehrenreich, Thibaud Metais, and Marcela Garita-Hernandez

Abstract

Patient-derived human-induced pluripotent stem cells (hiPSCs) are a valuable tool for validation of retinal gene therapies. These cells exhibit features identical to human embryonic stem cells (hESCs), such as self-renewal and pluripotency, and can differentiate into various retinal cell types, making them ideal for assessing clinically compliant viral vectors in a human retinal context. Several reprogramming methods have been developed, with integration-free methods being preferable due to their low risk of genomic integration and suitability for clinical applications. This chapter details the generation and characterization of patient-derived hiPSC lines from peripheral blood mononuclear cells (PBMCs) using non-integrative Sendai virus, with a focus on the thorough quality control (QC) necessary post-reprogramming. The QC includes confirming the expression of undifferentiated stem cell markers, the ability to differentiate into the three germ layers, and the absence of chromosomal abnormalities. This protocol provides a stepwise procedure for generating hiPSCs using an integration-free approach in 4 weeks.

Key words hiPSC, Patient-derived, PBMCs, Reprogramming, Embryoid body, Pluripotency, Chromosomal abnormalities

1 Introduction

The clinical relevance of human-induced pluripotent stem cell (hiPSC) research has paved a new path in the field of disease modeling. As translational medicine can transform benchtop discoveries into bedside therapies, developing accurate disease models is essential for understanding and treating diseases. This is particularly important when it comes to complex diseases that cannot be fully recapitulated using animal models alone [1, 2]. The most effective way to model genetic diseases is by recreating them in the very same environment where translational therapies will be applied: human tissues. However, not all tissues can be easily obtained from patients for study. The advent of hiPSCs resolves the ethical issues presented by embryonic stem cells (ESCs) [3] by

Livia S. Carvalho (ed.), *Retinal Gene Therapy: Methods and Protocols*, Methods in Molecular Biology, vol. 3023, https://doi.org/10.1007/978-1-0716-5198-8_9,
© The Author(s), under exclusive license to Springer Science+Business Media, LLC, part of Springer Nature 2026

reprogramming mature cells into pluripotent cells that can self-renew and differentiate into various other cell types [4].

hiPSCs have been derived from a range of sources, including human fibroblasts [5, 6], urine [7, 8], and blood [9, 10]. Blood cells offer a notable advantage as they can be collected in large quantities through noninvasive methods, unlike the dermal biopsy required to isolate fibroblasts [11]. Additionally, skin cells may demonstrate mosaicism due to prolonged UV radiation exposure [12]. Although various cellular reprogramming strategies have been developed, including retroviral vectors [13] and transposons [14], the non-integrative Sendai virus is often preferred due to its lack of insertional mutagenesis, high efficiency, and stable transgene expression [15–19].

Notably, patient-derived samples can be reprogrammed into hiPSC lines, allowing the development of personalized medicine. The differentiation of hiPSCs using 3D culture systems allows for the generation of retinal organoids [20–23]. These have been used for the development of modern cell therapies [20, 24–26], disease modeling [20, 21, 27–30], drug discovery [31, 32], screening of gene therapy [25, 33–35], and genome-editing tools [36–38].

Here, we described a detailed protocol for generating hiPSCs from patient PBMCs using non-integrative Sendai virus, under feeder-free conditions. This protocol also describes an extensive QC analysis and cell banking. The same QC analysis must be applied to healthy donor or genome-edited hiPSC control lines to ensure safety postediting.

2 Materials

2.1 Specialized Equipment

1. High-speed benchtop centrifuge up to 20,000 *g*.
2. Hemacytometer.
3. Fluorescent microscope.
4. Multi-application cell sorter such as the Sony MA900.
5. Bright-field microscope.
6. Gel box and electrophoresis system with gel combs.
7. DNA quantification such as the Nanodrop.
8. Benchtop thermocycler.
9. Real-Time PCR System such as the QuantStudio 7 Flex Real-Time PCR System with QuantStudio Real-Time PCR software (Thermo Fisher).
10. DNA sequence analysis software such as the SeqStudio Genetic Analyzer with GeneMapper software.
11. Cytogenetic drying chamber.

12. Gel Imaging System such as the ChemiDoc™ MP Imaging System with Image Lab™ software.
13. Cell freezing container.
14. Liquid nitrogen cell storage facilities.
15. ddPCR droplet generator with oil for probes.
16. ddPCR droplet reader.

2.2 Kits

1. CytoTune™-iPS 2.1 Sendai Reprogramming Kit.
2. iCS-digital™ PSC kit (Stem Genomics).
3. PowerPlex® or other STR DNA typing system.
4. RNA isolation kit.
5. cDNA synthesis kit.
6. DNA isolation kit.
7. Mycoplasma PCR Detection Kit.

2.3 Consumables

1. Six-well tissue culture-treated plate.
2. 12-well tissue culture-treated plates.
3. 24-well tissue culture-treated plate.
4. 24-well ultralow attachment tissue culture-treated plate.
5. Ultralow attachment tissue culture-treated 96-well plate.
6. 1.5 mL microcentrifuge tubes.
7. 15 mL conical tubes.
8. 50 mL conical tubes
9. Sepmate™ 50 mL tube.
10. Cryovials (2 mL).
11. 14 mL round-bottom tubes.
12. Parafilm™.
13. Circular glass coverslips (12 mm dia, #1).
14. Microscope slides (25 × 75 mm).
15. Needle (31 gauge, 8 mm).

2.4 Solutions

2.4.1 Cellular Reprogramming

1. Patient blood sample.
2. Ficoll-Paque density gradient medium.
3. 1X phosphate-buffered saline (PBS), pH 7.4: 800 mL deionized water, 8 g NaCl, 0.2 g KCl, 1.44 g, Na2HPO4, 0.24 g KH2PO4.
4. PBMC medium: 1X StemSpan SFEMII (Stem Cell Technologies), 1X erythroid expansion supplement (Stem Cell Technologies), 1% penicillin/streptomycin.
5. 0.05% Trypsin-EDTA (1X).

6. 1% Matrigel in 1X DMEM/F12.
7. ReproTeSR™ medium (Stem Cell Technologies).
8. mTeSR Plus™ medium (Stem Cell Technologies).
9. 0.5 mM EDTA in PBS.
10. ROCK Inhibitor/Y-27632 (10 μM).
11. ReLeSR™ (Stem Cell Technologies).
12. Freezing medium, such as Cryostor CS10.

2.4.2 Immuno-fluorescence Assay

1. iPSCs in 24-well plate.
2. 1X PBS.
3. 32% paraformaldehyde (PFA): water, 320 g paraformaldehyde powder, 1 N NaOH, HCl. For 1 L of 4% formaldehyde, add 800 mL of deionized water to a glass beaker on a stir plate in a ventilated hood. Heat while stirring to approximately 60 °C. Take care that the solution does not boil. Add 320 g of paraformaldehyde powder to the heated PBS solution. The powder will not immediately dissolve into solution. Slowly raise the pH by adding 1 N NaOH dropwise from a pipette until the solution clears. Once the paraformaldehyde is dissolved, the solution should be cooled and filtered. Adjust the volume to 1 L with 1X PBS. Recheck the pH and adjust it with small amounts of diluted HCl to approximately 6.9.
4. Washing buffer: 0.1% Tween-20, 1X PBS.
5. Permeabilization buffer: 0.1% Tween-20; 0.5% Triton X-100; 1X PBS.
6. Blocking buffer: 0.2% gelatin from cold water fish skin, 0.25% Triton X-100, 1X PBS (*see* **Note 1**).
7. Primary and secondary antibodies.
8. DAPI stain.
9. Ethanol (100%).
10. Mounting medium for coverslips.
11. Freezing medium, such as Cryostor CS10.

2.4.3 Embryoid Body Formation

1. iPSCs in a 24-well plate.
2. ReLeSR™ (Stem Cell Technologies).
3. mTeSR™ Plus medium (Stem Cell Technologies).
4. ROCK Inhibitor (10 μM).
5. Embryoid body medium I: DMEM/F12 GlutaMAX™ supplement, 20% KnockOut Serum Replacement (KOSR), 1X nonessential amino acids (NEAA), 1X penicillin/streptomycin, 0.1 mM 2-mercaptoethanol.

6. Embryoid body medium II: DMEM/F12 GlutaMAX™ supplement, 20% KnockOut Serum Replacement (KOSR), 10% Fetal Bovine Serum (FBS), 1X nonessential amino acids (NEAA), 1X penicillin/streptomycin, 0.1 mM 2-mercaptoethanol.
7. 1% Matrigel in 1X DMEM/F12.
8. 1X PBS.
9. 32% paraformaldehyde (PFA) (32%) in water.

2.4.4 T-PCT and qPCR

1. iPSCs in a six-well plate.
2. TaqMan™ Fast Advanced Master Mix.
3. 1X phosphate-buffered saline (PBS), pH 7.4.
4. ReLeSR™ (Stem Cell Technologies).
5. mTeSR Plus™ medium (Stem Cell Technologies).

2.4.5 Karyotyping

1. 1% ethidium bromide in water.
2. 1% colcemid in PBS.
3. 1X Trypsin-EDTA (0.5%).
4. 10% FBS solution in DMEM.
5. KCl (0.075 M).
6. Fixative mixture: 25% glacial acetic acid, 75% methanol.
7. Immersion mixture: 20% 10X Trypsin-EDTA (0.5%), 80% Hank's Balanced Salt Solution.
6. Staining mixture: Leishman's stain (2 g/L) in Gurr buffer.

2.4.6 Mycoplasma Detection Assay

1. 100 uL used cell culture media.
2. 1.5% agarose gel mixture: 98.5% TBE, 1.5% agarose, 0.001% SYBR Safe.
3. DNA loading dye.
4. DNA ladder.

2.4.7 Copy Variation Number (CNV) Analysis

1. ddPCR™ Supermix for Probes (No dUTP) (Bio-Rad).
2. HindIII-HF Enzyme.

2.5 Antibodies

Primary/secondary antibodies	Dilution used
OCT4	1:250
SOX2	1:400
SSEA4	1:400
NANOG	1:200

(continued)

Primary/secondary antibodies	Dilution used
SOX17	1:200
SMA	1:200
NESTIN	1:500
GFAP	1:200
Goat anti-mouse 488 secondary antibody	1:500
Goat anti-rabbit 555 secondary antibody	1:500
DAPI	1:1000

2.6 Primers

TaqMan™ probes
18S
NANOG
OCT4
ZFP42
TERT

3 Methods

All processes were performed at room temperature unless otherwise indicated.

3.1 Cellular Reprogramming

3.1.1 Isolation of PBMCs from Patient Samples (Fig. 1)

1. Pipette 15 mL of Ficoll-Paque density gradient medium through the central hole in a Sepmate™ 50 mL tube.
2. Dilute 8 mL of blood with 8 mL of PBS (dilution 1:1) in a separate 50 mL conical tube.
3. Carefully layer diluted blood over Ficoll-Paque density gradient medium. Pipette down the side of the 50 mL tube when adding diluted blood.
4. Centrifuge at 1200 *g* for 10 min at room temperature with the brake on.
5. Pour the upper supernatant from the Sepmate™ tube into a new 50 mL conical tube and discard the Sepmate™ 50 mL tube.
6. Add PBS to bring total volume to 50 mL.
7. Centrifuge at 400 *g* for 10 min.
8. Aspirate supernatant.

9. Resuspend the pellet in 10 mL of PBS and perform cell count manually using a Neubauer chamber.
10. Transfer 1 × 10^6 PBMCs into a 15 mL tube, centrifuge, and resuspend the pellet in 1 mL of PBMC medium. Plate into one well of a 12-well plate (no coating required, as monocytes are plastic-adherent). To ensure the cells are evenly spread in the well, gently shake the plate in a cross or a "figure 8" manner. Incubate in a 37 °C humidified incubator with 20% O_2 and 5% CO_2.
11. Spin down the remaining PBMCs at 400 *g* for 5 min.
12. Aspirate supernatant and resuspend pellet into freezing medium at a concentration between 2 × 10^6 and 4 × 10^6 cells per mL.
13. Transfer 1 mL of frozen cells into a labeled cryovial indicating the number of cells, the date of freezing, the passage number, and any other useful information.
14. Place at −80 °C overnight in a cell freezing container.
15. Transfer to liquid nitrogen tank the following day.

3.1.2 Transduction with Non-integrative Sendai Viral Vectors (Fig. 1)

1. Four days (day −4) before viral transduction, dissociate PBMCs with pre-warmed 1X Trypsin-EDTA (0.05%) for 5 min to detach cells from the plate. Then, perform cell count manually using a Neubauer chamber and transfer PBMCs to a 15 mL conical tube.
2. Centrifuge at 200 *g* for 10 min and resuspend in PBMC medium to bring the cell concentration to 5 × 10^5 cells per mL.
3. Add 1 mL of cell suspension per well to the center of at least four wells of a 24-well plate. Swirl the contents of the plate to ensure even spreading. Incubate in a 37 °C humidified incubator with 20% O_2 and 5% CO_2.
4. On days −3 to −1, perform half-media changes, replacing 0.5 mL PBMC medium from each well every day.
5. On day 0, perform cell count using a Neubauer chamber. Calculate the volume of virus required to obtain the desired multiplicity of infection (MOI) using the virus titer specific to each lot of KOS, hL-MYC, and hKLF4 virus provided in the CytoTune™-iPS 2.1 Sendai Reprogramming Kit. Recommended MOI for each of the three vectors is 5, 5, and 3, respectively.
6. For each transduction, transfer 2.5 × 10^5 total cells to a 14 mL round-bottom tube.
7. Add the previously calculated viral volume required for each of the three viral vectors to a 1 mL aliquot of 37 °C pre-warmed PBMC medium for each performed transfection. Pipette up

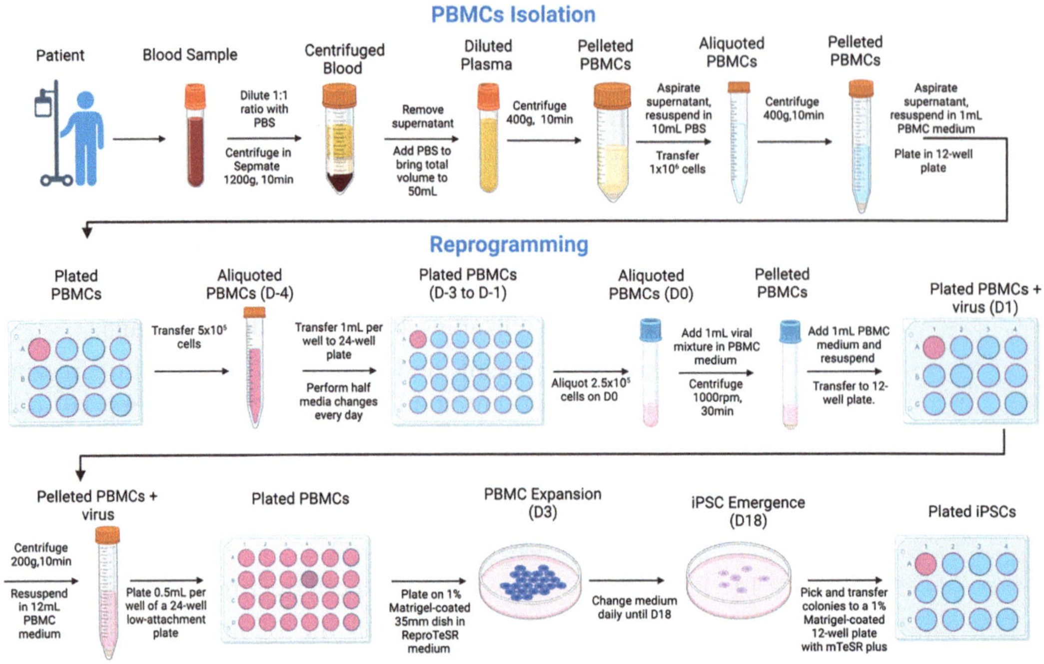

Fig. 1 A flowchart detailing the process of reprogramming patient-derived PBMCs into hiPSCs over the course of 3–4 weeks. First, a blood sample is taken from the patient. The PBMCs are isolated from whole blood via Ficoll density gradient medium and centrifugation and then subcultured. Next, non-integrative Sendai virus is used to deliver the KOS, hL-MYC, and hKLF4 reprogramming genes to the PBMCs. After transduction, the PBMCs are carefully maintained under conditions that will optimize reprogramming until hiPSC colonies emerge. These colonies are then manually picked, transferred, and expanded into a stable hiPSC line

and down to mix, and then transfer each 1 mL viral mixture to the 14 mL round-bottom tubes containing 2.5 × 10^5 cells prepared earlier.

8. Wrap the 14 mL round-bottom tube caps in Parafilm™ to create an airtight seal that prevents liquid loss and gas exchange, and then centrifuge at 1000 RPM for 30 min.
9. Add 1 mL PBMC medium to each tube, resuspend the cells, and then transfer each mixture to a well of a 12-well plate. Swirl the contents of the plate to ensure even spreading. Each mixture should be plated in a separate well of that plate. Incubate in a 37 °C humidified incubator with 20% O_2 and 5% CO_2.

3.1.3 Post-transduction Expansion and Maintenance (Fig. 1)

1. On day 1 (corresponding to the day after transduction), transfer the cells and medium to a 15 mL conical tube. Wash the well with 1 mL PBMC medium to ensure all cells have been transferred.
2. Centrifuge at 200 *g* for 10 min, aspirate supernatant, and resuspend in 12 mL PBMC medium.

3. Plate 0.5 mL of cell suspension into each well of a 24-well ultralow attachment plate. Incubate in a 37 °C humidified incubator with 20% O_2 and 5% CO_2. No media change is required for the next 2 days.
4. On day 3, seed PBMCs on a 1% Matrigel-coated 35 mm TC dish and maintain in ReproTeSR™ medium. Change media daily for 18 days.
5. On day 18, hiPSC colonies should begin to develop. Cultures are gradually changed to mTeSR Plus™ medium to support hiPSC colony formation. Media is refreshed daily until colonies emerge.
6. Once well-defined colonies (Fig. 2) have become visible (days 18–21), colonies can be picked using a 31-gauge, 8 mm needle. Use the needle to cut the hiPSC colonies in a grid-shaped pattern.
7. Using a 200 μL pipette tip, transfer the cut colonies to 1% Matrigel-coated 12-well plates for further expansion in mTeSR Plus™ medium (Fig. 2). Continue incubating at 37 °C with 20% O_2 and 5% CO_2.
8. Colonies are ready to be passaged when they reach approximately 85% confluency. Aspirate medium, and then wash the wells with PBS. Add 0.4 mL 0.5 mM EDTA in PBS per well. To ensure the EDTA fully coats the well, gently shake the plate doing a cross or a "figure 8" shape.
9. Incubate the plate at 37 °C for 5 min, and then aspirate the EDTA.

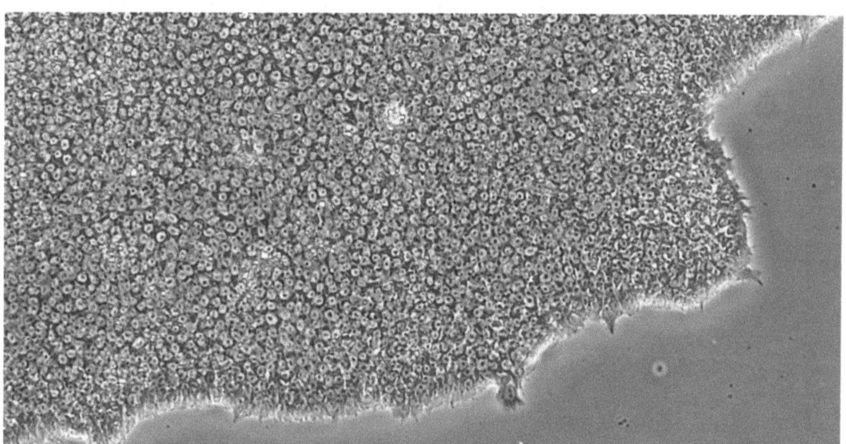

Fig. 2 hiPSC colony morphology. Notable features are as follows: a high nucleus-to-cytoplasm ratio within individual cells, spindle-shaped cell morphology, tightly packed cobblestone colony morphology, and a distinct colony border

10. Wash the wells with mTeSR Plus™ medium to dislodge cells, and then transfer them to 15 mL conical tube.
11. Plate cells on a new 1% Matrigel-coated 12-well plate. A 10 μM ROCK Inhibitor is added for 24 h to increase cell survival. Colonies are grown in a 37 °C humidified incubator with 20% O_2 and 5% CO_2. The cultures are routinely tested for *Mycoplasma* contamination using the Mycoplasma PCR Detection Kit as is described in Subheading 3.6 of this methods guide.
12. After the first five passages (5 weeks) of using this passaging method, the newly reprogrammed hiPSCs can be passaged using the following method, used for all iPSCs.

3.1.4 Passaging

1. Grow hiPSCs to 60–70% confluency in 1% Matrigel-coated six-well plate.
2. Wash each well with PBS (1X).
3. Add 1 mL of ReLeSR™.
4. Aspirate ReLeSR™ within 1 min and incubate the plate for 4 min at 37 °C.
5. Tap the plate to dislodge colony aggregates.
6. Resuspend cells in 2 mL mTeSR Plus™. Pipette up and down two to three times to reduce aggregate size. Optimal aggregates consist of small cell clusters, and single-cell homogenization should be avoided.
7. Passage the aggregates at 1/30 dilution once per week. This varies among hiPSC lines.
8. The remaining cells can be frozen or used for any immediate application.

3.1.5 Banking

1. After dissociating with ReLeSR™, transfer cell suspension from one well of a six-well plate to 15 mL conical tube and centrifuge for 4 min at 1000 RPM.
2. Aspirate supernatant and resuspend in 3 mL freezing medium.
3. Distribute 1 mL of the freezing medium + cell suspension mixture into three cryovials. Each cryovial should contain approximately 1.5×10^6 hiPCS.
4. Store cryovials at −80 °C overnight in a cell freezing container and then transfer to liquid nitrogen.

3.2 Pluripotency Marker Assay

3.2.1 Cell Culture and Plate Preparation

1. Grow hiPSCs to 60–70% confluency in a 24-well plate.
2. Wash each well with PBS.
3. Incubate each well with 4% paraformaldehyde aqueous (PFA) solution for 20 min at 4 °C.

4. Wash each well with 0.5 mL cold PBS three times to ensure no PFA remains in the plate.
5. Seal edges of plate with Parafilm™ and store at 4 °C until plate is ready for immunofluorescent staining.

3.2.2 Pluripotency Marker Immunofluorescent Staining and Imaging (Fig. 3)

1. Wash each well three times for 5 min each with PBS (1X).
2. Wash each well two times for 5 min each with washing buffer solution.
3. Wash each well with permeabilization buffer for 30 min to permeabilize the cells, allowing antibodies to enter and subsequently bind to their respective antigens.
4. Wash each well with blocking buffer for 30 min to minimize nonspecific binding of antibodies.
5. Prepare 200 μL of primary antibody solutions in blocking solution for each well (*see* **Notes 2** and **3**). Different antibodies required optimization. Our optimized antibodies and dilutions are as follows: OCT4 (1:250), SOX2 (1:400), SSEA4 (1:400), and NANOG (1:200) (*see* **Note 4**).
6. Add 200 μL aliquot of pluripotency marker antibody solution to each well (*see* **Note 5**). Incubate at 4 °C for 24 h.
7. Wash each well three times for 5 min each with washing buffer solution.
8. Prepare 200 μL of secondary antibody in blocking solution and add DAPI for nuclei counterstain for each well. Secondary antibodies are all diluted to 1:500 within the sample, and 1 μg/mL DAPI is diluted to 1:1000.
9. Add 200 μL of secondary antibody and DAPI solution to each well. Incubate at 37 °C for 30 min in the dark (*see* **Note 6**).
10. Wash each well three times for 5 min each with washing buffer solution.
11. Wash each well for 5 min with PBS (1X).
12. Wash each well for 1 min with ethanol (100%).
13. Add 30 μL of sample mounting medium to each well, and then place a coverslip over each well.
14. Store at 4 °C until imaging under fluorescent microscope (Fig. 3a).

3.2.3 Pluripotency Marker Immunostaining and Quantification via FACS Analysis

1. Grow hiPSCs to 60–70% confluency in a 24-well plate.
2. Wash each well twice with PBS (1X).
3. Add 1 mL of ReLeSR™.
4. Aspirate ReLeSR™ within 1 min and incubate the plate for 4 min at 37 °C.

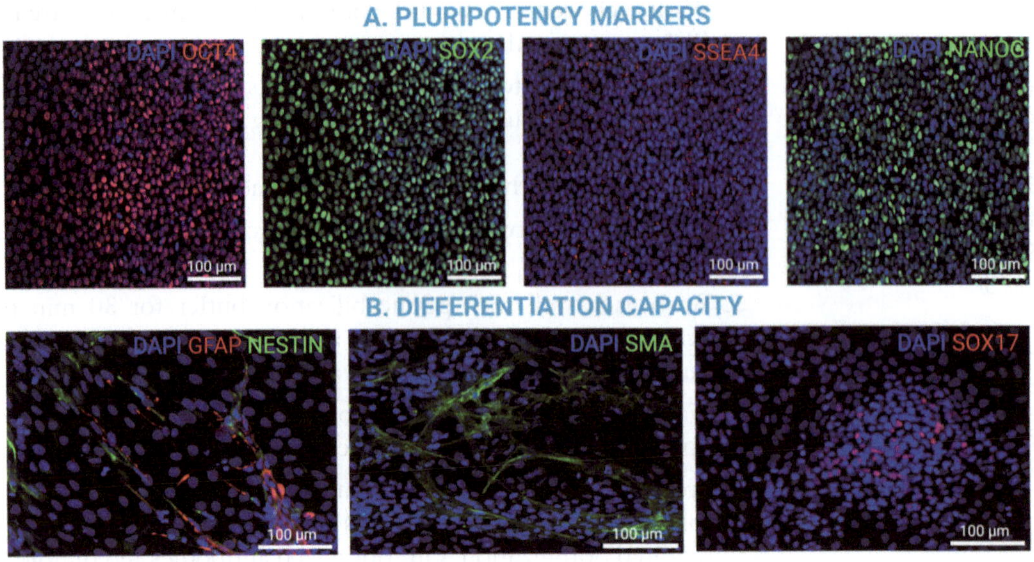

Fig. 3 Immunofluorescence analysis to confirm the presence of pluripotency markers (**a**) in hiPSCs and markers of differentiation capacity (**b**) in day 14 embryoid bodies. (**a**) hiPSC colonies expressed pluripotency markers OCT4, SOX2, SSEA4, and NANOG. (**b**) Embryoid bodies derived from hiPSC expressed markers of the three germ layers: endoderm, ectoderm, and mesoderm. Here, GFAP and NESTIN are stained to confirm the presence of ectodermal lineages, SMA is stained to confirm the presence of mesodermal lineages, and SOX17 is stained to confirm the presence of endodermal lineages. Scale 100 μm

5. Tap the plate to dislodge colony aggregates.
6. Wash suspended cells in cold PBS (1X), and centrifuge at 13000 RPM for 1 min using the benchtop centrifuge.
7. Fix suspended cells with cold 4% PFA in PBS for 20 min and centrifuge at 13000 RPM for 1 min.
8. Wash fixed cells twice with cold PBS (1X), and then centrifuge at 13000 RPM for 1 min.
9. Permeabilize suspended cells with permeabilization buffer for 10 min. Centrifuge at 13000 RPM for 1 min.
10. Add primary antibody solution to cell pellet. Incubate at 4 °C for 24 h.
11. Wash suspended cells three times with PBS (1X).
12. Incubate suspended cells with secondary antibody for 30 min at 37 °C.
13. Analyze at least 10,000 events in each experiment ($n = 3$) using a FACS sorter. Ensure that at least 80% of cells within the gated population return positive for staining.

3.3 Embryoid Body Assay

3.3.1 Spontaneous Differentiation/EB Generation

1. Grow hiPSCs to 90% confluency in a 24-well plate.
2. Incubate hiPSCs with ReLeSR™ (or comparable cell dissociation reagent) for 4 min at 37 °C.
3. Perform cell count manually using a Neubauer chamber.
4. Centrifuge cell suspension at 1000 RPM for 4 min.
5. Resuspend cells in mTeSR™ Plus with ROCK Inhibitor (10 μM) at a concentration of 1000 cells per 100 μL.
6. Seed 100 μL of cell suspension in each well of an ultralow attachment 96-well plate (*see* **Note 7**).
7. Allow embryoid bodies to aggregate in suspension culture for 3 days. During these 3 days, be careful to not move the plate or induce vibrations near the incubator, as it could impact the formation of EBs.
8. On day 3, change the growth medium to 200 μL of primary embryoid body medium. Perform half-medium changes every 2 days using fresh medium.
9. On day 7, transfer 16–24 embryoid bodies to a Matrigel-coated 24-well plate (*see* **Notes 8** and **9**). Grow in secondary embryoid body medium, performing half-medium changes every 2–3 days.
10. Aspirate medium on day 14 and wash cells with PBS.
11. Incubate each well with 4% paraformaldehyde aqueous (PFA) solution for 20 min at 4 °C.
12. Wash each well with 0.5 mL PBS.
13. Seal edges of plate with Parafilm™ and store at 4 °C until plate is ready for immunofluorescent staining.

3.3.2 Embryoid Body Immunofluorescent Staining/Imaging (Fig. 3)

1. Wash each well three times for 5 min each with PBS (1X).
2. Wash each well two times for 5 min each with washing buffer solution.
3. Wash each well with permeabilization solution for 30 min.
4. Wash each well with blocking solution for 30 min.
5. Prepare 500 μL primary antibody solution in blocking buffer for each well (*see* **Notes 2** and **3**). Our optimized antibodies and dilutions of choice are as follows: SOX17 (1:200), SMA (1: 200), NESTIN (1:500), and GFAP (1:200) (*see* **Note 10**).
6. Add 500 μL aliquot of pluripotency marker antibody + blocking solution to each well (*see* **Note 5**). Incubate at 4 °C for 24 h.
7. Wash each well three times for 5 min each with PBS-T solution.
8. Prepare 500 μL aliquots of secondary antibody with DAPI in blocking solution for each well.

Secondary antibodies are all diluted to 1:500 within the sample, and 1 μg/mL DAPI is diluted to 1:1000.

9. Add 500 μL of secondary antibody with DAPI solution to each well. Incubate at 37 °C for 30 min in the dark (*see* **Note 6**).
10. Wash each well three times for 5 min each with washing buffer solution.
11. Wash each well for 5 min with PBS (1X).
12. Wash each well for 1 min with ethanol (100%).
13. Add 30 μL of sample mounting medium to each well, and then place a coverslip over each well.
14. Store at 4 °C in the dark until imaging under fluorescent microscope (Fig. 3b).

3.4 RT-PCR and qPCR of Pluripotency Markers

3.4.1 RNA Extraction and Reverse Transcription

1. Grow cells to 70–90% confluency in one well of a six-well plate.
2. Wash each well twice with PBS (1X).
3. Add 1 mL of ReLeSR™.
4. Aspirate ReLeSR™ within 1 min and incubate the plate for 4 min at 37 °C.
5. Add 1 mL of mTeSR Plus medium.
6. Tap the plate to dislodge colony aggregates.
7. Transfer cell suspension to 15 mL conical tube. Centrifuge at 1000 RPM for 5 min to pellet cells, and then aspirate supernatant. Freeze at −20 °C if not proceeding directly to next step.
8. Isolate RNA using RNA extraction kit.
9. Quantify extracted RNA using Nanodrop.
10. Perform reverse transcription of RNA into cDNA using cDNA synthesis kit.

3.4.2 qPCR, Data Analysis, and Statistics

1. Perform 1:5 dilution of cDNA in water.
2. Add 5 μL TaqMan™ Fast Advanced Master Mix for each 10 μL sample, and then separately add sample-specific DNA and triplicate-specific probe for each pluripotency marker (*see* **Notes 11** and **12**).
3. Run qPCR, and then use ΔΔCT method to find fold induction value.
4. Normalize target mRNA levels to reference gene 18S. Use statistical software to identify if fold change of pluripotency markers is statistically significant in comparison.

3.5 Karyotyping

3.5.1 G-Banded Karyotyping Assay

1. Grow culture to at least 45% confluency (*see* **Note 13**).
2. Incubate cells with ethidium bromide (1%) for 5 min to prevent DNA condensation and elongate chromosomes.

3. Incubate cells with colcemid (10 μL/mL) for 5 min to prevent spindle formation during mitosis, resulting in metaphase arrest.
4. Incubate cells with pre-warmed 1X Trypsin-EDTA (0.5%) for 5 min to detach cells from plate (*see* **Note 14**).
5. Transfer dissociated cell suspension to 10% FBS solution and centrifuge at 300 *g* for 10 min.
6. Aspirate supernatant and resuspend in 0.075 M KCl, and then incubate for 5 min.
7. Invert tube, and then add fixative mixture. Invert again, and then centrifuge at 300 *g* for 10 min. Aspirate supernatant and repeat this step twice with fresh fixative mixture each time, incubating the cells in fixative mixture for 5 min each time they are resuspended.
8. Drop cell suspension onto labeled microscope slides (*see* **Note 15**). Allow slides to dry.
9. Age slides at 90 °C for 60 min in cytogenic drying chamber or at room temperature overnight.
10. Dip aged slides in immersion mixture.
11. Rinse slide in running water, and then immerse in staining mixture and agitate.
12. Rinse slide in filtered water and dry.

3.5.2 Karyotyping and Chromosomal Abnormality Analysis

1. Check appearance of chromosomes under bright-field microscope.
2. Perform chromosome count on at least 20 cells, analyze chromosomes on a minimum of eight cells, and create karyograms for a minimum of four cells.
3. Ensure counted cells are normal diploid (46 chromosomes) and that two of each chromosomal homolog are identified.
4. Align chromosome homologues for karyogramming.
5. Take note of any structural abnormalities by comparing homologues to each other and to reference images (*see* **Note 16**) and include them in the karyogram (Fig. 4).

3.6 Mycoplasma Detection Assay

3.6.1 Media Acquisition and PCR

1. Remove 100 μL of media when performing routine media change. Media should be taken from one well of all hiPSC-related plates in culture, including those of hiPSCs and EBs.
2. Using the Mycoplasma PCR Detection Kit, generate master mix for $((n \times 2) + 3)$ wells, where $n =$ number of samples. This allows all samples to be run in duplicate alongside one positive and one negative control (*see* **Note 19**). For each reaction tube, use 12.5 μL PCR master mix, 1 μL primer mix, 2.5 μL media sample, and 9 μL dH2O for a final reaction volume of 25 μL (*see* **Note 17**).

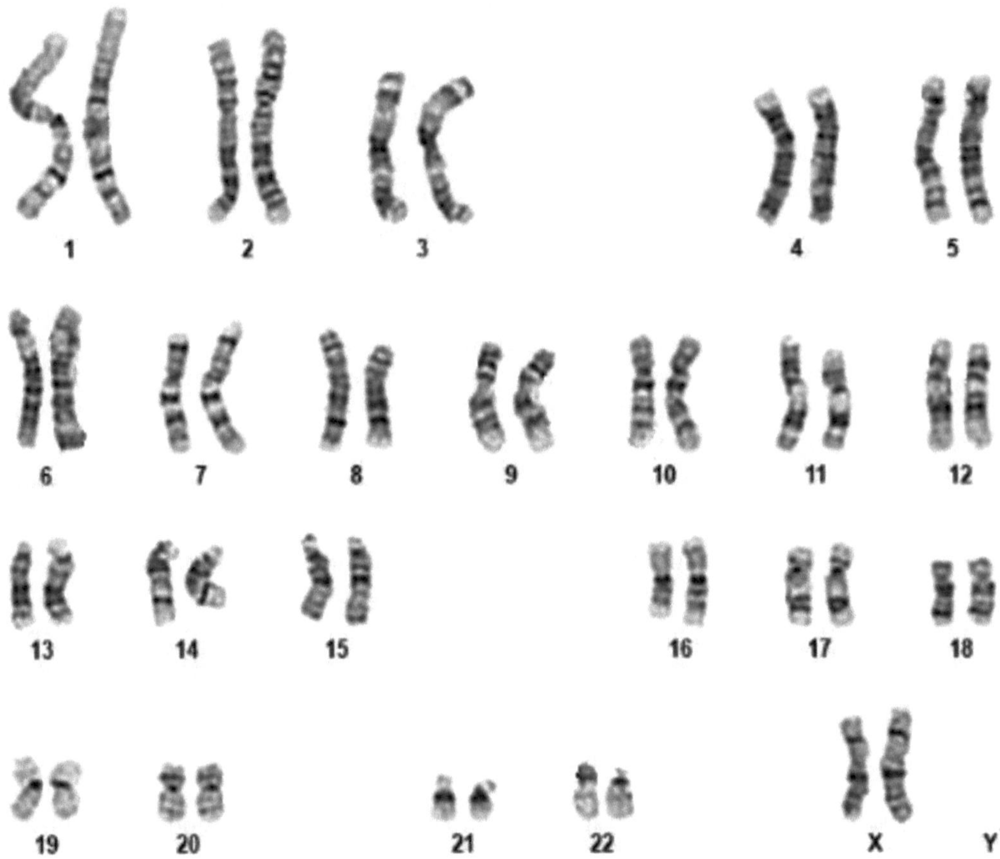

Fig. 4 G-banded karyogram of 23 chromosomal pairs in a human female. Note the light and dark regions within each chromosome pair. These appear when heterochromatin is stained along the chromosome arms. The resulting bands vary in size and location, allowing for homologous chromosomes to be paired together and isolated from other chromosomes. These signature banding patterns also allow for detection of chromosomal aberrations in the event that large parts of the chromosome are deleted or duplicated

3.6.2 Agarose Gel Preparation and Electrophoresis

1. Prepare the 1.5% agarose gel mixture in an Erlenmeyer flask. Microwave the TBE and agarose in the Erlenmeyer flask, stopping frequently to swirl the flask to help dissolve the agarose. Once solution appears fully clear to the eye, add SYBR Safe at a 1X final concentration.
2. Pour the 1.5% agarose gel mixture into a gel case and add a comb that will form as many wells as is necessary for the assay (*see* **Note 18**). Account for all samples in duplicate, one positive control, one negative control, and a DNA ladder. Wait for the gel to solidify, and then carefully remove the comb, ensuring well integrity is not compromised.

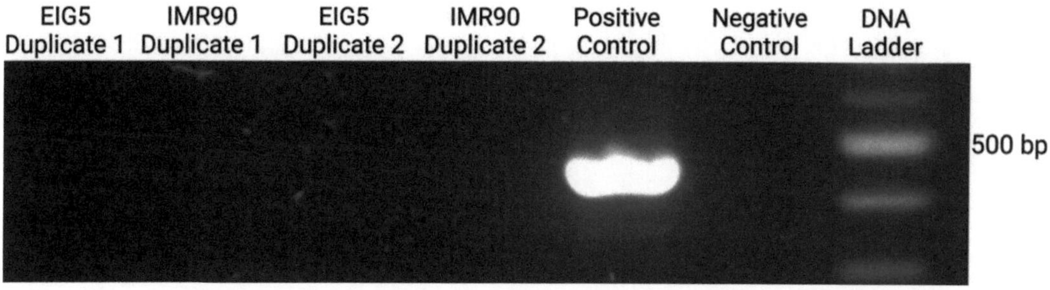

Fig. 5 *Mycoplasma* test by PCR analysis. Each cell culture (EIG5 and IMR90) is run in duplicate alongside a positive control (yielding one band at approximately 500 bp), a negative control, and a 100 bp DNA ladder for reference. No bands are seen in any lanes other than that of the positive control, indicating that the two samples are free of *Mycoplasma* contamination

3. Load 10 μL of each PCR product into the gel wells, remembering to include the positive and negative controls, as well as a 100 bp DNA ladder.
4. Run the gel at 120 V for 60 min.
5. Image the gel using the ChemiDoc™ MP Imaging System.
6. Check the gel to ensure that a strong band is present in the positive control lane and that no bands other than faint primer dimers are present in any sample lanes (Fig. 5).

3.7 Short Tandem Repeat (STR) DNA Fingerprinting

3.7.1 DNA Isolation and Preparation

1. Isolate DNA using DNA isolation kit.
2. Quantify extracted DNA using Nanodrop and dilute to 0.67 ng/μL.
3. Use PowerPlex 16 HS System to amplify DNA loci containing STRs for analysis.

3.7.2 Capillary Electrophoresis

1. Prepare loading cocktail for capillary electrophoresis using PowerPlex kit loading mixture.
2. Perform capillary electrophoresis for fragment analysis on amplified sample (*see* **Note 20**).

3.7.3 Electrophoresis Data Analysis

1. Use GeneMapper™ Software to determine if any peaks differentiate the sample from the positive control. Positive control peaks should not differ from the sample peaks, and no peaks should be present in the negative control (*see* **Notes 21** and **22**).
2. If deviations are present, sample can be rerun to see if these are STR artifacts or genuine alleles.
3. If over 80% of STR sequences are identical matches between sample and positive control, sample is declared to have passed.

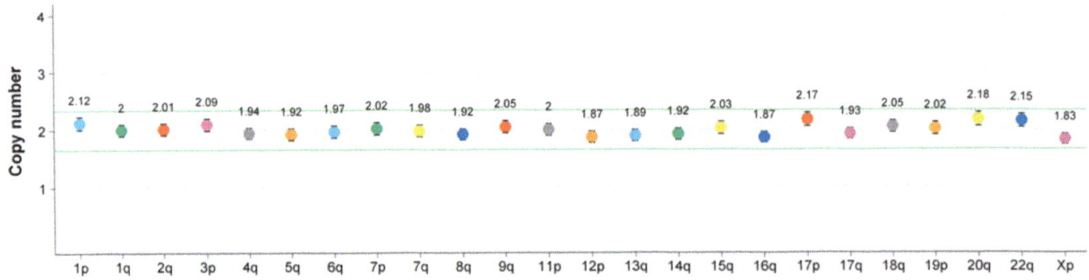

Fig. 6 Copy number variation analysis for hiPSCs derived from a human female patient. The graph depicted illustrates the test results for the 24 most common copy number variations within hiPSC lines. These 24 loci constitute over 90% of CNVs found in hiPSC lines. Note that all 24 values fall between the upper and lower acceptable limits (demarcated using two green bars). A range is permitted due to the presence of some few cells in a population possessing these CNVs. However, because the CNVs are within the acceptable range of 2 (as a normal diploid human female will have two copies of every test locus), the fraction of hiPSCs containing CNVs is not significant enough to warrant concern for the genomic integrity of the line, and the test is considered passed

3.8 Copy Number Variation (CNV) Analysis

3.8.1 DNA Isolation and Preparation

1. Isolate DNA using DNA isolation kit.
2. Quantify extracted DNA using Nanodrop.

3.8.2 ddPCR

1. Using iCS-digital™ PSC kit, generate reaction mixtures of 150 ng of purified genomic DNA, 1x ddPCR™ Supermix for Probes, 900 nM of each primer, and 250 nM of each probe.
2. Load reaction mixes into cartridges with 70 μL of droplet generation oil per sample.
3. Generate droplets using droplet generator.

3.8.3 PCR and Plate Reading

1. Transfer emulsions to a thermocycler and amplify according to the following parameters: one stage of 95 °C for 10 min, 45 cycles of 95 °C for 30 s and 60 °C for 60 s, and then one stage of 98 °C for 10 min.
2. Load plate onto droplet reader and analyze using QuantaSoft software. If copy number falls within acceptable range for all 24 chromosomal loci analyzed, the test is passed (Fig. 6).

4 Notes

1. Vortexing PBS + gelatin alone is optimal to ensuring all gelatin is dissolved within PBS before adding Triton-X. Mixing after adding Triton-X results in significant bubble formation within the bottle. Use a glass bottle with screw lid to prevent reagent spillage during vortexing process.

2. This aliquot size is specific to a plate size of 24 wells. It is important to note that this volume can be adjusted to account for variable plate sizes, which can make this procedure more cost-effective.
3. It is highly recommended to create master mixes of $n + 1$ for each antibody combination when generating these aliquots to minimize labor.
4. These are our chosen antibodies, though there are other options for pluripotency markers, such as TRA-1-60.
5. It is highly recommended to perform each stain in duplicate due to the length of this procedure.
6. All subsequent steps and handling procedures (even those beyond this assay, such as imaging) should be performed in the dark. This can be accomplished by keeping the plate wrapped in tinfoil at all times to prevent light exposure from disrupting secondary antibody fluorescence.
7. Use multichannel pipette to seed 96-well plate in order to minimize time expenditure.
8. Multiple EBs can be grown in the same well, though care should be taken to ensure that they are not crowding each other and can unfurl properly as they develop.
9. Transfer pipettes will work well for extracting the embryoid bodies from the wells of the 96-well plate and replating them in the 24-well plate without damaging them in the process.
10. These antibodies were our chosen antibodies, though there are many other options for germ layer markers.
11. The pluripotency markers we probe are NANOG, OCT4, ZFP, and TERT, all of which are compared against 18S expression as a housekeeping gene. These can be changed if another pluripotency marker is desired.
12. Pluripotency markers should be tested in triplicate to ensure consistency.
13. Colonies at peak mitotic activity (3–5 days post-passage) work best for this assay.
14. Pipette up and down to dissociate cells and break up chunks if necessary.
15. Coat the slide with fixative, using a paper towel to drain excess.
16. If normal and abnormal cells are present, karyogram a minimum of three abnormal cells and one normal cell.
17. Make sure to perform all controls in tandem with any kits used. Positive and negative controls are extremely important for this assay.
18. Be extremely careful of bubble formation when pouring the gel. Bubbles can interfere with sample migration by disrupting

the structure of the gel matrix. If bubbles form during the gel pouring, a pipette tip can be used to manually pop the bubbles and prevent these issues. This must be performed prior to gel solidification.

19. For example, if three samples are to be tested for *Mycoplasma* contamination, prepare a master mix for nine wells: six sample wells (two for each of the three original samples), one positive control well, one negative control well, and one extra well volume to account for any pipetting error that may occur throughout the process.
20. Recommended to leave reaction plate in electrophoresis machine until it has been determined reinjection is not required or the 24-hour window has passed.
21. A true peak is determined by examining each peak by size and shape to confirm correct peak appearance.
22. Each sample should show a "Primer Flash" or a cluster of peaks from all dye channels between 1000 and 1500 that exceeds 32,000 RFUs.

References

1. Seifirad S (2014) An emerging need for developing new models for myocardial infarction as a chronic complex disease: lessons learnt from animal vs human studies on cardioprotective effects of Erythropoietin in reperfused myocardium. Front Physiol 5
2. Seifirad S, Haghpanah V (2019) Inappropriate modeling of chronic and complex disorders: How to reconsider the approach in the context of predictive, preventive and personalized medicine, and translational medicine. EPMA J 10: 195
3. Charitos IA, Ballini A, Cantore S et al (2021) Stem cells: a historical review about biological, religious, and ethical issues. Stem Cells Int
4. Takahashi K, Yamanaka S (2006) Induction of pluripotent stem cells from mouse embryonic and adult fibroblast cultures by defined factors. Cell 126:663
5. Nakagawa M, Koyanagi M, Tanabe K et al (2008) Generation of induced pluripotent stem cells without Myc from mouse and human fibroblasts. Nat Biotechnol 26:101
6. Montserrat N, Nivet E, Sancho-Martinez I et al (2013) Reprogramming of human fibroblasts to pluripotency with lineage specifiers. Cell Stem Cell 13:341
7. Kim EY, Page P, Dellefave-Castillo LM et al (2016) Direct reprogramming of urine-derived cells with inducible MyoD for modeling human muscle disease. Skelet Muscle 6
8. Zhou T, Benda C, Duzinger S, et al (2011) Generation of induced pluripotent stem cells from urine 22
9. Loh YH, Hartung O, Li H et al (2010) Reprogramming of T cells from human peripheral blood. Cell Stem Cell 7:15
10. Staerk J, Dawlaty MM, Gao Q et al (2010) Reprogramming of human peripheral blood cells to induced pluripotent stem cells. Cell Stem Cell 7:20
11. Park IH, Zhao R, West JA et al (2008) Reprogramming of human somatic cells to pluripotency with defined factors. Nature 451:141
12. Abyzov A, Mariani J, Palejev D et al (2012) Somatic copy number mosaicism in human skin revealed by induced pluripotent stem cells. Nature 492:438
13. Hotta A, Ellis J (2008), Retroviral vector silencing during iPS cell induction: an epigenetic beacon that signals distinct pluripotent states
14. Yusa K, Rad R, Takeda J et al (2009) Generation of transgene-free induced pluripotent mouse stem cells by the piggyBac transposon. Nat Methods 6:363
15. Yang W (2014) iPSC reprogramming from human peripheral blood using Sendai virus mediated gene transfer. Stembook
16. Ban H, Nishishita N, Fusaki N et al (2011) Efficient generation of transgene-free human induced pluripotent stem cells (iPSCs) by

temperature-sensitive Sendai virus vectors. Proc Natl Acad Sci USA 108:14234
17. Fusaki N, Ban H, Nishiyama A et al (2009) Efficient induction of transgene-free human pluripotent stem cells using a vector based on Sendai virus, an RNA virus that does not integrate into the host genome. Proc Jpn Acad Ser B Phys Biol Sci 85:348
18. Hu K (2014) Vectorology and factor delivery in induced pluripotent stem cell reprogramming. Stem Cells Dev 23:1301
19. Ye L, Muench MO, Fusaki N et al (2013) Blood cell-derived induced pluripotent stem cells free of reprogramming factors generated by sendai viral vectors. Stem Cells Transl Med 2:558
20. Meyer JS, Shearer RL, Capowski EE et al (2009) Modeling early retinal development with human embryonic and induced pluripotent stem cells. Proc Natl Acad Sci USA 106: 16698
21. Reichman S, Terray A, Slembrouck A et al (2014) From confluent human iPS cells to self-forming neural retina and retinal pigmented epithelium. Proc Natl Acad Sci U S A 111: 8518
22. Zhong X, Gutierrez C, Xue T et al (2014) Generation of three-dimensional retinal tissue with functional photoreceptors from human iPSCs. Nat Commun 5
23. Zhang X, Wang W, Jin ZB (2021) Retinal organoids as models for development and diseases. Cell Regen 10:33
24. Garita-Hernandez M, Guibbal L, Toualbi L et al (2018) Optogenetic light sensors in human retinal organoids. Front Neurosci 12
25. Gonzalez-Cordero A, Kruczek K, Naeem A et al (2017) Recapitulation of human retinal development from human pluripotent stem cells generates transplantable populations of cone photoreceptors. Stem Cell Reports 9:820
26. Sharma R, Khristov V, Rising A et al (2019) Clinical-grade stem cell-derived retinal pigment epithelium patch rescues retinal degeneration in rodents and pigs. Sci Transl Med 11
27. Lane A, Jovanovic K, Shortall C et al (2020) Modeling and rescue of RP2 retinitis Pigmentosa using iPSC-derived retinal organoids. Stem Cell Reports 15:67
28. Sanjurjo-Soriano C, Jimenez-Medina C, Erkilic N, et al (2023) USH2A variants causing retinitis pigmentosa or Usher syndrome provoke differential retinal phenotypes in disease-specific organoids. 4, Hum Genet Genomics Adv
29. Atkinson R, Georgiou M, Yang C et al (2024) PRPF8-mediated dysregulation of hBrr2 helicase disrupts human spliceosome kinetics and 5′-splice-site selection causing tissue-specific defects. Nat Commun 15:3138
30. Rodrigues A, Slembrouck-Brec A, Nanteau C et al (2022) Modeling PRPF31 retinitis pigmentosa using retinal pigment epithelium and organoids combined with gene augmentation rescue. NPJ Regen Med 7:39
31. Vergara MN, Flores-Bellver M, Aparicio-Domingo S, et al (2017) Three-dimensional automated reporter quantification (3D-ARQ) technology enables quantitative screening in retinal organoids. 144
32. Sharma R, George A, Nimmagadda M et al (2021) Epithelial phenotype restoring drugs suppress macular degeneration phenotypes in an iPSC model. Nat Commun 12:7293
33. Burnight ER, Gupta M, Wiley LA, et al (2017) Using CRISPR-Cas9 to generate gene-corrected autologous iPSCs for the treatment of inherited retinal degeneration. 25
34. Garita-Hernandez M, Routet F, Guibbal L et al (2020) AAV-mediated gene delivery to 3D retinal organoids derived from human induced pluripotent stem cells. Int J Mol Sci 21
35. Khabou H, Garita-Hernandez M, Chaffiol A et al (2018) Noninvasive gene delivery to foveal cones for vision restoration. JCI Insight 3
36. Garita-Hernandez M, Zeaiter M, Pierce EA (2023) Optimized prime editing efficiently generates mutations in hiPSCs to model inherited retinal degeneration. Invest Ophthalmol Vis Sci 64:750–750
37. Galdikaite-Braziene E, Serrano JCC, Garita-Hernandez M et al (2023) Characterization of EYS in human iPSC-derived retinal organoids. Invest Ophthalmol Vis Sci 64:3186–3186
38. Brydon EM, Bronstein R, Buskin A et al (2019) AAV-mediated gene augmentation therapy restores critical functions in mutant PRPF31+/− iPSC-derived RPE cells. Mol Ther Methods Clin Dev 15

Chapter 10

Transduction of AAV Vectors in iPSC-Derived Retinal Organoids

Deborah Aubin and Anai Gonzalez-Cordero

Abstract

Stem cell-derived organoids research provides an exciting model and technology to develop new therapies for difficult and intractable diseases. Human-induced pluripotent stem cells (hiPSCs) have the potential to differentiate into organ-like three-dimensional structures, especially when it is challenging to obtain patient samples for this purpose. This chapter outlines a comprehensive differentiation protocol for the generation and manipulation of retinal organoids derived from pluripotent stem cells (PSCs), including embryonic stem cells (ESCs) and hiPSCs. The method involves culturing PSCs, differentiating them into retinal organoids, and in vitro transduction of organoids with adeno-associated virus (AAV) vectors. AAVs can be used to label a desired population, to test novel promoters, and to test gene therapies. We will also describe key steps following organoid transduction, which entail collection, immunostaining, imaging and preservation for molecular biology readouts. The protocol provides a detailed guide for researchers working in the field of retinal biology and gene therapy.

Key words Pluripotent stem cells, Differentiation, Retinal organoids, Adeno-associated virus (AAV) vectors, Transduction, Gene therapy

1 Introduction

Numerous differentiation protocols have been established for generating retinal organoids, offering promising tools for various research applications. In this chapter, we will focus on a 2D/3D protocol adapted from the work of Reichman et al. (2014) [1] and Zhong et al. (2014) [2]. These protocols have been further improved through the addition of lipids and glucose supplementation, resulting in enhanced differentiation and specifically the generation of photoreceptor cell outer segments, a critical milestone typically observed from 22 weeks onward [3]. This enhancement in photoreceptor development facilitates the maturation of retinal organoids, offering a valuable tool for disease modeling.

Livia S. Carvalho (ed.), *Retinal Gene Therapy: Methods and Protocols*, Methods in Molecular Biology, vol. 3023,
https://doi.org/10.1007/978-1-0716-5198-8_10,
© The Author(s), under exclusive license to Springer Science+Business Media, LLC, part of Springer Nature 2026

Comparative transcriptome studies have been conducted on 30-weeks-old retinal organoids, adult and fetal retinal tissue, and organoids generated by different protocols [4]. These investigations have revealed striking similarities between the cell types present in organoids and those found in the in vivo retina, affirming the authenticity of these retinal organoids as reliable replicas of human retinas.

The credibility of these bona fide retinal organoids makes them excellent models for testing novel therapies, such as AAV gene therapy. Prior studies have already demonstrated the effectiveness of using these organoids to model inherited retinal degenerations (reviewed in: 5, 6) and to establish the potency of in vitro gene therapy in addressing various eye disorders [3, 6, 7].

Additionally, retinal organoids serve as a versatile platform for studying gene expression regulation. Researchers can employ them to test new promoters and evaluate their specificity to certain cell types, notably in the case of cone and rod photoreceptor cells, as demonstrated by Gonzalez-Cordero et al. (2018) [8] and others. This capability aids in fine-tuning gene therapy approaches for maximum effectiveness.

Furthermore, retinal organoids are valuable tools for assessing the tropism of novel AAV capsids [8–10]. This is particularly important in gene therapy, where the efficiency and specificity of viral vectors play a pivotal role in the success of treatment strategies. The use of retinal organoids as a testing ground for these vectors can accelerate the development of safer and more effective gene therapy approaches for retinal diseases and beyond.

In summary, this section provides a detailed explanation of the methodology to differentiate hiPSC-derived retinal organoids, as well as the procedures to transduce retinal organoids in vitro using AAV gene therapy. The major procedures covered include stem cell differentiation and transduction techniques, followed by collection methodologies for various experimentations including immunostaining, RNA/DNA/protein isolation, and dissociation of organoids for procedures such as single-cell RNA sequencing (ScRNA) and cell transplantation.

2 Materials

2.1 Retinal Organoid Differentiation

1. E8 media.
2. E6 media.
3. Proneural induction media (PIM): Advanced Dulbecco's Modified Eagle Medium (A-DMEM)/F12 (500 mL), Minimum Essential Medium Non-Essential—Amino Acids solution 100X (MEM-NEAA) (1X), N2 Supplement 100X (1X), glutamine 200 mM (1.9 mM), penicillin/streptomycin (Pen/Strep, 10,000 U/mL) (96 U/mL), amphotericin B (0.5 μg/mL).

4. Retina differentiation media (RDM): Dulbecco's Modified Eagle Medium, high glucose (DMEM) (350 mL), Ham's F-12 Nutrient Mix GlutaMAX™ Supplement (F12) (150 mL), B27 minus vitamin A (4 mM), penicillin/streptomycin (Pen/Strep, 10,000 U/mL) (192 U/mL), amphotericin B (0.5 μg/mL).
5. Retina differentiation media + factors (RDMF): Dulbecco's Modified Eagle Medium, high glucose (DMEM) (315 mL), Ham's F-12 Nutrient Mix GlutaMAX™ Supplement (F12) (135 mL), B27 minus vitamin A (4 mM), heat-inactivated fetal bovine serum (Hi-FBS) (50 mL; *see* **Note 1**), taurine (0.1 mM), GlutaMAX™ Supplement (4 mM), penicillin/streptomycin (Pen/Strep, 10,000 U/mL) (188 U/mL), amphotericin B (0.5 μg/mL).
6. Maturation media (ALT70): Advanced DMEM/F12 (A-DMEM/F12) (450 mL), heat-inactivated fetal bovine serum (Hi-FBS) (50 mL; *see* **Note 1**), GlutaMAX™ Supplement (4 mM), B27 minus vitamin A (4 mM), Taurine (0.1 mM) penicillin/streptomycin (Pen/Strep, 10,000 U/mL) (188 U/mL), amphotericin B (0.5 μg/mL), retinoic acid (RA working solution, 1 uM; *see* **Note 2**).
7. Maturation media and maintenance (ALT90): Advanced DMEM/F12 (A-DMEM/F12) (500 mL), GlutaMAX™ Supplement (4 mM), B27 minus vitamin A (4 mM), N2 Supplement 100X (1X), glucose solution (0.7 M), Taurine (0.1 mM), chemically defined lipids 1000X (1X), penicillin/streptomycin (Pen/Strep, 10,000 U/mL) (184 U/mL), amphotericin B (0.9 μg/mL), retinoic acid (RA working solution, 0.5 uM; *see* **Note 2**).
8. CO2 incubator: Required for maintaining organoid cultures.
9. Low-binding 96-well plates: Used for 3D early cultures of retinal organoids.
10. Low-binding 24-well plates: Used for long-term cultures of retinal organoids.

2.2 Immunohistochemistry

1. 4% paraformaldehyde (PFA): Dulbecco's phosphate-buffered saline (DPBS), 40 mL, 16% PFA 10 ml. Essential for fixation of retinal organoids.
2. 20% sucrose solution: Sucrose powder (10 g), Dulbecco's phosphate-buffered saline (DPBS), 50 mL (*see* **Note 3**).
3. OCT compound: Used for embedding retinal organoids.
4. Liquid nitrogen: Essential for rapid freezing of embedded organoids.
5. 2-Methylbutane: Used in conjunction with liquid nitrogen for cryopreservation.

6. Positively charged glass slides: Necessary for mounting and staining of cryosections.
7. Hydrophobic pen: Used for drawing a hydrophobic barrier during immunostaining.
8. Secondary antibodies: Required for specific immunostaining (conjugated with fluorophores or enzymes).
9. Nuclear counterstain (e.g., DAPI): Optional for visualizing cell nuclei.
10. Mounting medium: Used to mount and preserve stained sections on glass slides.

2.3 Virus Transduction

1. AAV vectors: Adeno-associated virus vectors used for transduction of retinal organoids.
2. ALT90 organoid culture medium: A specialized medium used as the base for diluting AAV viral vectors and for culturing organoids during and after transduction.
3. Low-binding 24-well plates: Used for culturing and transducing retinal organoids.
4. Sterile 1.5 mL Eppendorf tubes: For handling vector-containing medium.

Please note that the specific AAV vectors used in this procedure should be selected based on the target gene and the experimental requirements. We use $1.13E+10^{11}$ vg/organoid.

The concentration of AAV vectors should be determined according to the desired transduction efficiency and specific experimental goals.

2.4 DNA, RNA, and Protein Extraction

1. *DNA Isolation:*
   - DNA isolation kit: Used for DNA isolation.
   - Spectrophotometer: Used for measuring DNA concentration and purity.
2. *RNA Isolation:*
   - RNA isolation kit: Used for RNA isolation.
   - β-Mercaptoethanol (β-ME): Added to Buffer RLT for RNAse deactivation.
   - Spectrophotometer: Used for measuring RNA concentration and purity.
3. *Protein Isolation:*
   - RIPA buffer: Used for total protein extraction from organoids.
   - Low protein binding tubes: Used for collecting organoids for protein isolation.

- 4 °C or ice: Required for performing protein isolation steps at low temperature.
- Bradford, BCA, or Lowry assay reagents: Used for measuring protein concentration.
- Sterile microcentrifuge tubes. Used for collection, processing and storage of protein extracts.

2.5 Organoid Dissociation

1. Wide-bore P1000 pipette tips: Used for collecting organoids without damaging it.
2. Sterile laminar safety cabinet: Required for maintaining aseptic conditions during the dissociation procedure.
3. Neurosphere dissociation kit (P): Kit components used for organoid dissociation.
4. Please note that specific reagents and kits for organoid dissociation may include Enzyme P (papain), Enzyme A (DNase), Buffer Y, and Buffer X as mentioned later in the text.
5. Sterile 1.5 mL and 2 mL microcentrifuge tubes: Used for handling cells and organoids during dissociation.
6. Cell counting chamber: Used for cell counting, if necessary.

3 Methods

3.1 Retinal Organoid Differentiation

1. Maintain PSCs as previously described [11] until they reach 90–95% confluence (*see* **Note 4**).
2. At day 0 of differentiation, replace E8 or any PSC media with 2 mL of warm E6 media for 2 days (D1 and D2).
3. On day 3, add 2 mL of proneural induction media (PIM) to induce neural differentiation for up to 7 weeks (*see* **Note 5**).
4. Lightly pigmented islands of retinal pigmented epithelium (RPE) should appear around week 3, and neuroretinal optic vesicles will appear in close proximity to these RPE regions between weeks 4 and 7 in culture.
5. From week 4 onward, neuroretinal vesicles should be manually excised using an EVOS dissection microscope inside the safety cabinet. Two 21G needles can be used for this procedure.
6. Dissected single organoids are transferred to low-binding 96-well plates and cultured individually in 120 μL of retinal differentiation media (RDM). Media is replaced with fresh 120 μL of RDM until day 42.
7. At 6 weeks of differentiation, RDM is supplemented with 10% FBS, 100 uM taurine, and 2 mM GlutaMAX and 120 μL added to each well.

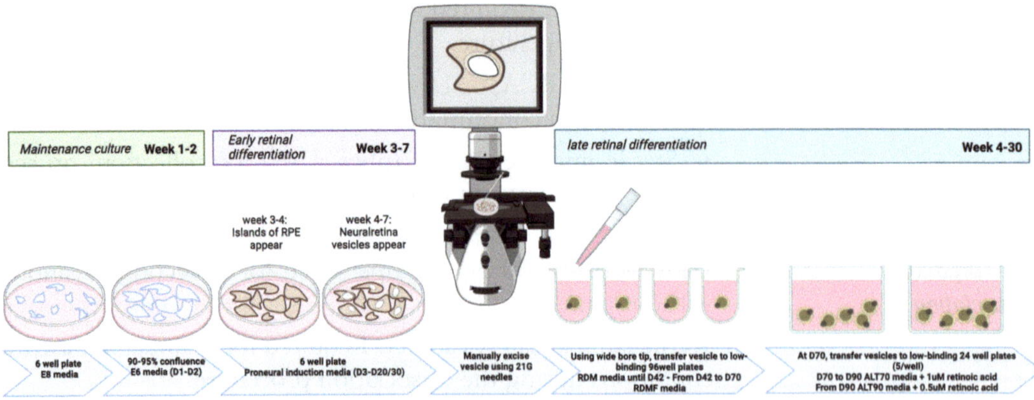

Fig. 1 Schematic of retinal organoid differentiation from pluripotent stem cells

8. At 10 weeks, vesicles are transferred to low-binding 24-well plates (five vesicles/well) in 1 mL of ALT70 media supplemented with 1 uM retinoic acid (RA). Change media every other day with 1 mL of fresh ALT70.
9. At 12 weeks, media is changed to ALT90 (ALT70 supplemented with 1% N2), and RA concentration is reduced to 0.5 uM. Change media every other day with 1 mL of fresh ALT90. A schematic diagram of the process is shown in Fig. 1.

3.2 Retinal Organoid AAV Transduction

1. Treat retinal organoids with AAV vectors between 15 and 19 weeks of differentiation to target specifically photoreceptor cells (*see* **Notes 6** and 7; Fig. 2).
2. Dilute the desired volume of AAV vectors in 300 μL of fresh ALT90 medium (300 μL/well of 24-well plate) to achieve the desired multiplicity of infection (MOI) or concentration.
3. Aspirate ALT90 medium from the wells containing between two and five organoids.
4. Add 300 μL of the vector-containing medium to each well containing organoids to transduce.
5. Tilt the plate in the incubator so that the retinal organoids are submerged by media containing the virus.
6. The following day (first thing in the morning), top up the treated wells with 700 μL of ALT90 medium and incubate for 72 h in media containing virus.
7. After the incubation period, carefully aspirate the vector-containing medium and replace it with 1 mL of fresh ALT90 medium.
8. Incubate the transduced organoids for the period necessary for your experiment (see notes above).
9. Monitor the organoids for changes in gene expression, protein levels, or any other relevant outcomes.

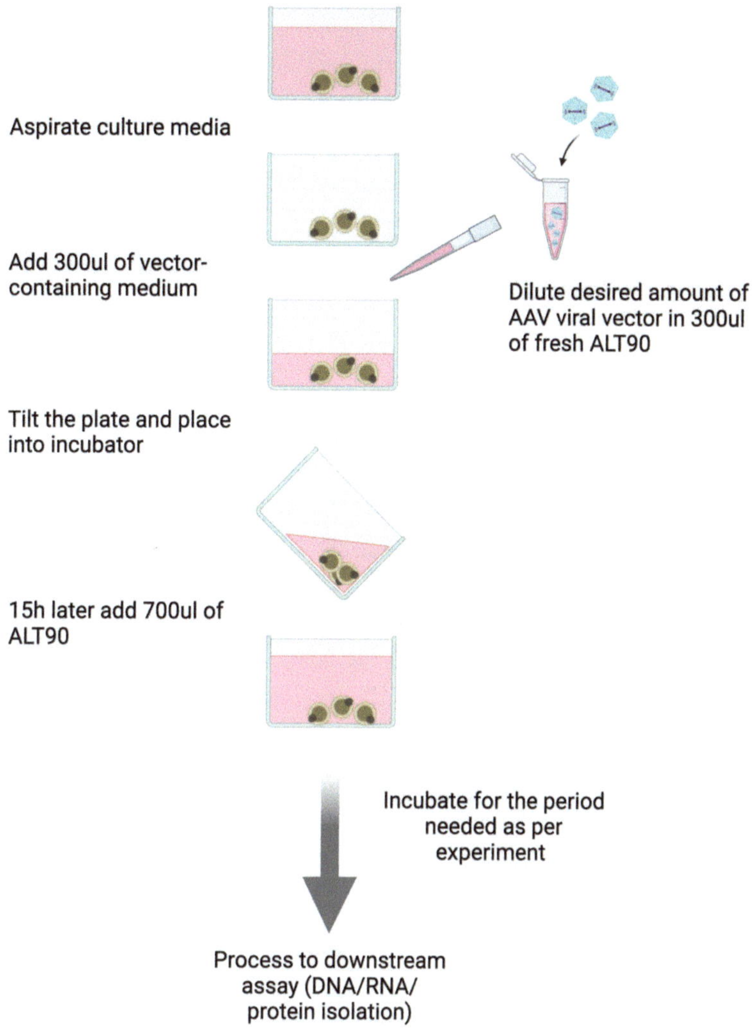

Fig. 2 Step-by-step AAV transduction protocol of retinal organoids

3.3 Organoid Collection for Molecular Biology (Fig. 3)

3.3.1 Snap Freezing

1. Place liquid nitrogen in a safe benchtop HDPE Dewar container.
2. Collect organoids under the safety cabinet from 24-well plates using a wide-bore p1000 tip and place it into 1.5 mL Eppendorf tube (*see* **Note 8**).
3. Remove the culture medium from 1.5 mL Eppendorf tube carefully and wash the organoids once with PBS.
4. Remove PBS from the tube and directly place the tube into liquid nitrogen using crucible tongs. Ensure that the freezing process is as rapid as possible to minimize ice crystal formation.

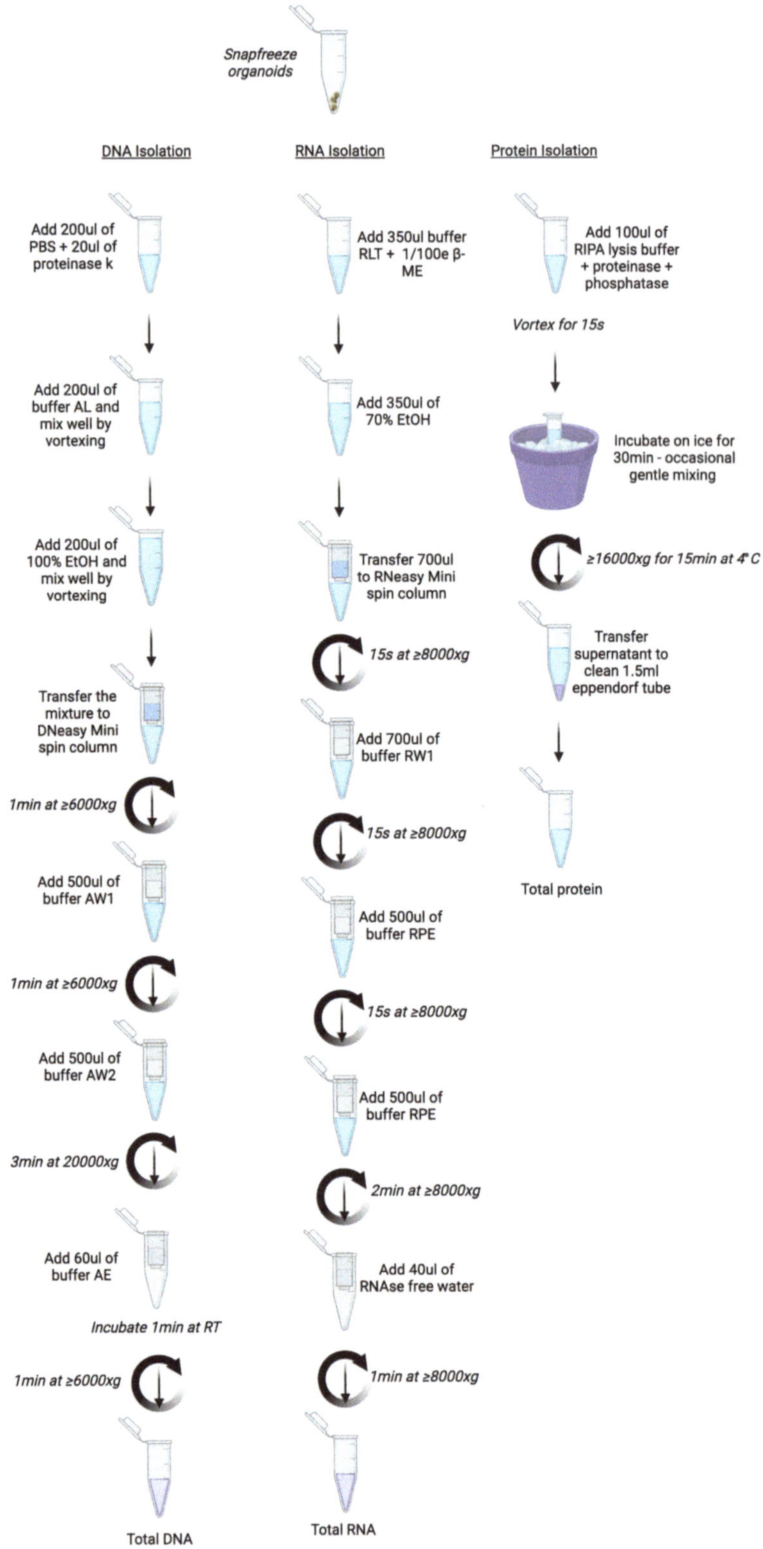

Fig. 3 Procedure overview for DNA/RNA and protein isolation from snap-frozenretinal organoids

5. Once organoids are completely frozen, transfer the tube to a long-term storage container (e.g., a labeled box) and store them at −80 °C freezer.

3.3.2 DNA Isolation

1. Collect organoids from −80 °C freezer and place it on ice (*see* **Note 9**).
2. Resuspend frozen organoids in 200 μL PBS and add 20 μL proteinase K (provided in the kit).
3. To lyse organoids, add 200 μL Buffer AL (provided in the kit). Mix thoroughly by vortexing for about 30 s.
4. Add 200 μL of 100% ethanol (molecular biology grade). Mix thoroughly by vortexing for about 30 s.
5. Pipette the mixture into a DNeasy Mini spin column (provided in the kit) placed in a 2 mL collection tube.
6. Centrifuge at ≥6000× *g* (8000 rpm) for 1 min. Discard the flow-through and collection tube.
7. Place the spin column in a new 2 mL collection tube. Add 500 μL Buffer AW1 (provided in the kit). Centrifuge for 1 min at ≥6000× *g*. Discard the flow-through and collection tube.
8. Place the spin column in a new 2 mL collection tube, add 500 μL Buffer AW2 (provided in the kit), and centrifuge for 3 min at 20,000× *g* (14,000 rpm). Discard the flow-through and collection tube.
9. Transfer the spin column to a new 1.5 mL or 2 mL Eppendorf tube.
10. Elute the DNA by adding 60 μL Buffer AE (provided in the kit) to the center of the spin column membrane. Incubate for 1 min at room temperature (15–25 °C). Centrifuge for 1 min at ≥6000× *g*.

    Optional: Repeat step 8 for increased DNA yield.
11. Measure the DNA concentration and purity using a spectrophotometer.
12. Store the isolated DNA at −20 °C for downstream analysis such as targeted region amplification using PCR, followed by Sanger sequencing.

3.3.3 RNA Isolation

1. Collect organoids from −80 °C freezer and place it on ice (*see* **Note 10**).
2. To lyse organoids, add 350 μL of Buffer RLT (provided by kit; *see* **Note 11**) into the organoids containing tube and pipette up and down with a p1000 first to disrupt organoids and then with P200 to completely lyse the organoid cells.
3. Add 350 μL of 70% ethanol to the lysate and mix well by pipetting. Do not vortex or centrifuge.

4. Transfer up to 700 μL of the sample, including any precipitate, to an RNeasy Mini spin column placed in a 2 mL collection tube (provided by kit).
5. Close the lid, and centrifuge for 15 s at ≥8000× *g*. Discard the flow-through.
6. Add 700 μL of Buffer RW1 (provided by kit) to the RNeasy spin column. Close the lid, and centrifuge for 15 s at ≥8000× *g*. Discard the flow-through.
7. Add 500 μL of Buffer RPE (provided by kit) to the RNeasy spin column. Close the lid, and centrifuge for 15 s at ≥8000× *g*. Discard the flow-through.
8. Add 500 μL of Buffer RPE (provided by kit) to the RNeasy spin column. Close the lid, and centrifuge for 2 min at ≥8000× *g*.
9. Place the RNeasy spin column in a new 2 mL collection tube (supplied). Centrifuge at full speed for 1 min to dry the membrane.
10. Place the RNeasy spin column in a new 1.5 mL collection tube (supplied). Add 40 μL RNase-free water directly to the spin column membrane. Close the lid, and centrifuge for 1 min at ≥8000× *g* to elute the RNA.
11. Measure the RNA concentration and purity using a spectrophotometer.
12. Store the isolated RNA at −80 °C for long-term storage for downstream analysis such as RT-PCR or RT-qPCR.

3.3.4 Protein Isolation

1. Collect organoids from −80 °C freezer and place it on ice (*see* **Note 12**).
2. To lyse organoids, add 100 μL of ice-cold RIPA buffer supplemented with protease and phosphatase inhibitors to the organoids containing tube.
3. Vortex the tube vigorously for 15 s to ensure thorough mixing.
4. Incubate the tube on ice for 30 min with occasional gentle mixing.
5. Centrifuge the lysate at high speed (e.g., 12,000× *g*) for 15 min at 4 °C to pellet cell debris and insoluble material.
6. Carefully transfer the supernatant (containing the protein extract) to a clean microcentrifuge tube.
7. Measure the protein concentration using a suitable protein assay method (e.g., Bradford, BCA, or Lowry).
8. Adjust the concentration with Milli-Q water as needed for downstream applications.
9. Store the protein extract at −80 °C for long-term storage or proceed with immediate downstream applications.

3.4 Organoid Collection for Immunolabeling

1. Collect retinal organoids from in vitro cultures in 1.5 mL Eppendorf tubes.
2. Gently wash them once with PBS to remove residual media.
3. Remove PBS and fix organoids in ~500 μL 4% paraformaldehyde (PFA) at room temperature for 45 min then wash once with PBS.
4. Remove PBS and add ~500 μL of 20% sucrose in PBS solution to the tubes containing organoids (*see* **Note 13**).
5. Store at 4 °C in the fridge until use.
6. For embedding, position retinal organoids in the center of cryo-mold using a wide-bore tip and make sure that you remove the sucrose completely.
7. Add enough OCT compound to cover the organoids and make sure they are positioned again to center using a 20 μL pipette tip. Add more OCT to fill the cryo-mold.
8. Fill up liquid nitrogen in a safe benchtop HDPE Dewar container and place it under a laminar flow cabinet.
9. Place a container (Becher) with 50 mL of 2-methylbutane (isopentane) into the Dewar container. Allow the 2-methylbutane to cool until it becomes very cold but not frozen.
10. Transfer the cryo-molds rapidly into 2-methylbutane containing Becher in the liquid nitrogen. Ensure they are fully submerged.
11. Once the OCT in the cryo-molds turn white, pull them out of 2-methylbutane using crucible tongs.
12. Store them in −20 °C freezer until cryosectioning.

3.4.1 Cryosectioning

1. Set the cryostat chamber to −20 °C.
2. Securely place the OCT-embedded organoid block on the microtome stage within the cryostat.
3. Align and advance microtome blades to cut thin sections of the organoid (*see* **Note 14**).
4. Transfer cryosections adhering to the microtome blade onto positively charged glass slides.
5. Allow sections to adhere to the slides for at least 20 min.
6. Store cut slides at −20 °C until used for immunostaining.

3.4.2 Immunostaining

1. Remove slides from −20 °C storage and let them thaw at room temperature for about 10 min while keeping them in a humidified chamber to prevent drying.
2. Draw a barrier using hydrophobic pen.

3. Wash slides in PBS (pH 7.4) for 5 min to remove residual OCT compound and rehydrate the tissue.
4. Incubate slides in blocking solution (1% BSA, 0.1% Triton X100, and 5% serum from the host species of your secondary antibody) for 1–2 h at room temperature to block nonspecific binding sites.
5. Dilute primary antibodies in blocking solution (minus serum) with the recommended manufacturer's concentration. Prepare a volume of 200–300 μL solution per slide.
6. Remove blocking solution and add 200–300 μL diluted primary antibody solution to the tissue sections on the slides.
7. Incubate slides in a humidified chamber at 4 °C overnight (or according to the recommended incubation time provided by the antibody manufacturer) to allow primary antibodies to bind specifically to target proteins.
8. Wash slides three times with PBS for 5 min each to remove excess primary antibody.
9. Dilute appropriate secondary antibodies (conjugated with fluorophores or enzymes) in blocking solution as per the manufacturer's recommendations.
10. Apply the diluted secondary antibody solution to the tissue sections.
11. Incubate slides in a dark, humidified chamber at room temperature for 2 h only (or as recommended by the antibody manufacturer).
12. Wash slides three times with PBS for 5 min each to remove excess secondary antibody.
13. If using a nuclear counterstain (e.g., DAPI), apply it to the slides as per the manufacturer's instructions to visualize cell nuclei.
14. Apply a small drop of mounting medium to each tissue section.
15. Gently place a coverslip on top to seal the sections.
16. Allow the mounting medium to set by storing the slides in the dark at room temperature for 10 min. Seal corners of the coverslip with drops of nail polish.

3.5 Organoid Dissociation

Organoid dissociation is a critical step for downstream applications such as single-cell analysis or passaging. Follow these steps carefully to ensure successful dissociation.

1. Prepare dissociation solution using the components of the neurosphere dissociation kit (P) as below (*see* **Notes 15** and **16**):
   (a) 25 μL Enzyme P (papain).
   (b) 5 μL Enzyme A (DNase).

   (c) 10 μL Buffer Y.
   (d) 960 μL Buffer X.
   (e) Complete dissociation solution 1 mL.
2. Add 250 μL of dissociation solution for up to 10 organoids.
3. Briefly warm complete dissociation solution in 37 °C water bath (approx. Guide time is 2 min per 1 mL).
4. Collect 10–20 organoids using a wide-bore p1000 tip in 1.5 mL Eppendorf tubes.
5. Aspirate media in each tube and wash organoids with 1 mL of PBS.
6. Aspirate PBS and add 250 μL of dissociation solution in the tube. Ensure that the organoids are fully covered with the dissociation solution.
7. Incubate in 37 °C water bath for 5 min.
8. After 5 min, gently agitate organoids by flicking the tubes (*see* **Note 17**).
9. Incubate for a further 5 minutes in 37 °C water bath.
10. Using a p1000 pipette, gently triturate each tube of organoids (approx. Ten times). Tissue should begin to disrupt and break into small clumps.
11. Incubate for a further 5 min in 37 °C water bath. Usually, a total of 15 min at 37 °C incubation is sufficient for proper dissociation, but timing can be extended.
12. To dissociate the organoids further, switch to a p200 tip and pipette up and down approx. Ten times. The solution should now be clear and with little to no visible clumps of tissue.
13. To deactivate dissociation solution, add 1 mL of HBSS buffer (for 250 μL of enzyme solution).
14. Filter the cell solution using Miltenyi 30 μm pre-separation filters (130-041-407) into new Eppendorf tubes.
15. Wash the dissociation tube with 500 μL of HBSS to collect any remaining cells.
16. Wash the filter with 500 μL of HBSS to collet cells that remained in the filter.
17. Pellet cells with centrifugation at 900 RPM for 5 min.

    Optional: Count cells using countess. Collect 10 μL of cell suspension (+ 10 μL trypan blue for counting) and place solution into countess chamber.
18. After centrifugation, aspirate media and resuspend cells in medium or buffer for downstream applications. The choice of medium or buffer will depend on the specific experiments or analyses you plan to perform (*see* **Note 18**).

4 Notes

1. Take the defrosted fetal bovine serum (FBS) bottle and submerge it in a water bath set at 56 °C, making sure the water covers the bottle up to slightly above the serum level. Begin a timer for 30 min. Periodically gently swirl the serum every 5–10 min to ensure even heating and prevent protein coagulation at the bottom of the bottle.
2. Retinoic acid (RA) stocks and working solution are light sensitive. Dilute 50 mg of RA into 1.7 mL of DMSO—make aliquots and store at −20 °C. Once thawed, RA stock aliquot can be keep at 4 °C for 1 week. On feeding days, prepare a RA working solution: A90 media + RA at 1/1000 dilution factor.
3. Agitate until sucrose is dissolved, and then filter-sterilize using 30 μm filter. Uncontaminated solution can be stored at 4 °C indefinitely.
4. PSCs to be cultured in six-well plates.
5. Feed differentiation cultures every Monday, Wednesday, and Friday until the end of retinal organoid culture.
6. Other cell types can be targeted earlier; however, proliferative cells are present until week 10: Transduction before this time point is not recommended.
7. For titers, a range of 1 to $5E+10^{11}$ vg/organoid can be tested without organoid toxicity. Titer will depend on construct and experimental goal. Two weeks following transduction should be sufficient to check for reporter marker expression and tropism of novel capsids. For gene therapy, including gene editing with CRISPR/Cas9, organoids should be analyzed starting at 4 weeks. Transduction should be performed in the late afternoon so that organoids are not maintained in only 300 μL of media for too long.
8. Snap-freezing in liquid nitrogen is a critical step for preserving cellular integrity and biological molecular content for downstream analyses. Organoid number to be frozen will depend on the downstream analyses to perform.
9. For genomic DNA, snap-freeze at least two organoids together to obtain a concentration of around 80 ng/μL. The rest of this protocol will refer to the buffers associated with the QIAGEN DNeasy blood and tissue kit but can be substituted with the equivalent buffers/reagents from other equivalent kits.
10. For RNA isolation, snap-freeze at least two organoids together to obtain a concentration of around 45 ng/μL. The rest of this protocol will refer to the buffers associated with the QIAGEN RNeasy mini kit but can be substituted with the equivalent buffers/reagents from other equivalent kits.

11. Add 10 μL β-mercaptoethanol (β-ME, not provided by kit) to 1 mL Buffer RLT to aid in RNAse deactivation.
12. Ensure all reagents are kept on ice and perform all steps at 4 °C or on ice. For mass spectrometer proteome analysis, 20 organoids/condition is needed. Organoids must be collected into low protein binding tube. For Western blots, three to five organoids are recommended obtaining around 2 μg/μL, and a total of 20 μg of protein is sufficient for Western blot assay.
13. 20% sucrose is used to avoid ice crystal formation. Once in sucrose, organoids will float at the surface. Store in 4 °C fridge for as long as needed. Before embedding, ensure that organoids settle at the bottom of the tube which confirms proper sucrose infiltration and protection.
14. For retinal organoids, 14 μm sections in thickness are recommended. An average of six slides can be used to collect a good representation of the sample.
15. Ensure aseptic technique under the safety cabinet throughout the procedure, and work quickly to minimize cell stress. Adapt the procedure as needed for specific organoid types and sizes. This protocol outlines the basic steps for dissociating hiPSC-derived organoids using the neurosphere dissociation kit (P) from Miltenyi, but any other neurosphere dissociation kit can be used instead.
16. Preparation of enzyme buffer: Do not vortex DNAse containing solutions as DNAse is denatured by vortexing.
17. When flicking tubes during incubation in water bath, always ensure organoids collect at the tube base and that they are not adhering to the walls.
18. These dissociation incubation guidelines are suitable for week 17 "good" batches of organoids. Older organoids, or those with "fluffy" or large RPE regions, may require extended incubation (15 min max) post-trituration. Some contaminating brain, non-retinal tissue, or large RPE clumps may never dissociate completely. These will later be excluded from solution after filtering.

References

1. Reichman S, Terray A, Slembrouck A, Nanteau C, Orieux G, Habeler W, Nandrot EF, Sahel J-A, Monville C, Goureau O (2014) From confluent human IPS cells to self-forming neural retina and retinal pigmented epithelium. Proc Natl Acad Sci USA 111(23): 8518–8523. https://doi.org/10.1073/pnas.1324212111
2. Zhong X, Gutierrez C, Xue T, Hampton C, Vergara MN, Cao L-H, Peters A et al (2014) Generation of three dimensional retinal tissue with functional photoreceptors from human IPSCs. *Nat Commun* 5:4047. https://doi.org/10.1038/ncomms5047
3. West EL, Majumder P, Naeem A, Fernando M, O'Hara-Wright M, Lanning E, Kloc M et al (2022) Antioxidant and lipid supplementation

improve the development of photoreceptor outer segments in pluripotent stem cell-derived retinal organoids. Stem Cell Reports 17(4): 775–788. https://doi.org/10.1016/j.stemcr.2022.02.019

4. Jieun HK, O'Hara-Wright M, Kim D, Loi TH, Lim BY, Jamieson RV, Gonzalez-Cordero A, Yang P (2023) Comprehensive characterization of fetal and mature retinal cell identity to assess the Fidelity of retinal organoids. Stem Cell Reports 18(1):175–189. https://doi.org/10.1016/j.stemcr.2022.12.002
5. Andreazzoli M, Barravecchia I, De Cesari C, Angeloni D, Demontis GC (2021) Inducible pluripotent stem cells to model and treat inherited degenerative diseases of the outer retina: 3D-organoids limitations and bioengineering solutions. Cells 10(9):2489. https://doi.org/10.3390/cells10092489
6. Kamil K, Swaroop A (2020) Pluripotent stem cell-derived retinal organoids for disease modeling and development of therapies. Stem Cells 38(10):1206–1215. https://doi.org/10.1002/stem.3239
7. Kamil K, Qu Z, Gentry J, Fadl BR, Gieser L, Hiriyanna S, Batz Z et al (2021) Gene therapy of dominant CRX-Leber congenital Amaurosis using patient stem cell-derived retinal organoids. Stem Cell Rep 16(2):252–263. https://doi.org/10.1016/j.stemcr.2020.12.018
8. Gonzalez-Cordero A, Goh D, Kruczek K, Naeem A, Fernando M, Kleine Holthaus S-M, Takaaki M et al (2018) Assessment of AAV vector tropisms for mouse and human pluripotent stem cell–derived RPE and photoreceptor cells. Hum Gene Ther 29(10):1124–1139. https://doi.org/10.1089/hum.2018.027
9. McClements ME, Steward H, Atkin W, Archer Goode E, Gándara C, Chichagova V, MacLaren RE (2022) Tropism of AAV vectors in photoreceptor-like cells of human IPSC-derived retinal organoids. Transl Vis Sci Technol 11(4):3. https://doi.org/10.1167/tvst.11.4.3
10. Garita-Hernandez M, Routet F, Guibbal L, Khabou H, Toualbi L, Riancho L, Reichman S et al (2020) AAV-mediated gene delivery to 3D retinal organoids derived from human induced pluripotent stem cells. Int J Mol Sci 21(3):994. https://doi.org/10.3390/ijms21030994
11. Gonzalez-Cordero A, Kruczek K, Naeem A, Fernando M, Kloc M, Ribeiro J, Goh D et al (2017) Recapitulation of human retinal development from human pluripotent stem cells generates transplantable populations of cone photoreceptors. Stem Cell Reports 9(3): 820–837. https://doi.org/10.1016/j.stemcr.2017.07.022

Part III

Protocols for In Vivo Proof-of-Concept Animal Studies and Therapy Validation

Chapter 11

Intravitreal Injections in Mouse Models of Eye Disease

Annie L. Miller, Yashvi Bhatt, and Livia S. Carvalho

Abstract

Intravitreal injections are a commonly used technique in the treatment of some vitreoretinal diseases. In clinical situations, intravitreal injections are often used to deliver drugs that treat diseases such as age-related macular degeneration, retinal vein occlusion, and uveitis. Many preclinical studies in animals such as mice utilize intravitreal injections to test neuroprotective treatments and gene therapies before clinical use can be indicated. Herein, we describe our method and favorable practices for intravitreal injection into the eye of both preweaned and adult mice.

Key words Intravitreal injection, Mouse models, Drug delivery, Transscleral, Vitreoretinal disease

1 Introduction

Intravitreal injections allow the delivery of therapeutic agents to the vitreous humor of the eye. It is a commonly performed ophthalmic surgery, second only to cataract surgery [1]. Intravitreal injections are used to treat various posterior eye segment diseases, for example, age-related macular degeneration, retinal vein occlusion, and uveitis, among others [2–4]. They are a generally well-tolerated procedure among patients with relative safety, although they are known to increase the risk of infections and retinal detachment and can cause transient changes in intraocular pressure [1]. Intravitreal injections are also commonly used in preclinical studies, for example, in the context of delivering neuroprotective agents or gene therapies to the retina in retinal degenerative diseases [5, 6]. The use of intravitreal injections in preclinical studies poses an interesting challenge due to differences between most animal eyes and human eyes. This paper focuses on intravitreal injections in the mouse eye due to their extensive use in ophthalmic research. The main challenges that arise when performing intravitreal injections in mice come from the small size of the eye and the comparatively large size of the lens. As the mouse eye is drastically smaller than a

Livia S. Carvalho (ed.), *Retinal Gene Therapy: Methods and Protocols*, Methods in Molecular Biology, vol. 3023,
https://doi.org/10.1007/978-1-0716-5198-8_11,
© The Author(s), under exclusive license to Springer Science+Business Media, LLC, part of Springer Nature 2026

human eye, the volume administered to the vitreous is limited to an absolute maximum of 2 μL, with lower volumes generally suggested to avoid retinal swelling, changes to intraocular pressure, and potential drug reflux [7–9]. Additionally, the mouse lens makes up a substantial portion of the vitreous, meaning that the operator has to navigate the vitreous cavity while avoiding damaging the lens and causing cataracts as a side effect of the procedure [1]. In this paper, we describe best practice for intravitreal administration in both preweaned and adult mice, with the injection through the pars plana using a transscleral approach, which helps to avoid damaging the lens.

2 Materials

2.1 Drugs and Solutions Required for the Procedure

1. Injectable anesthesia: 80 mg/kg ketamine, 10 mg/kg xylazine. To make the anesthesia mixture for adult mice, we add 0.8 mL of ketamine from a stock concentration of 100 mg/mL and 0.5 mL of xylazine from a stock solution of 20 mg/mL. The ketamine and xylazine is mixed with 3.7 mL of 0.9% sterile saline. We suggest the use of injectable anesthesia over inhaled anesthesia (*see* **Note 1**). We then weigh animals before anesthetization, using the following equation to calculate how much anesthesia they require: *anesthesia volume (mL) = 0.5*weight (g)/100*. For example, for a 15 g mouse, the calculation would be (0.5*15 g)/100 = 0.075 mL of anesthesia mix.
2. Injectable anesthesia reversing agent: 1 mg/kg atipamezole. To make the anesthesia reversal agent for adult mice, mix together 4.9 mL of 0.9% sterile saline and 0.1 mL of atipamezole from a stock concentration of 5 mg/mL. The reversal volume that is applied to the mouse follows this formula: *reversal volume (mL) = weight (g)/100*. For example, for a 15 g mouse, the calculation would be 15 g/100 = 0.15 mL of anesthesia reversal mix.
3. Pupil dilating drops: 5 mg/mL (0.5%) tropicamide eye drops. Dilating the pupils of adult mice is recommended before intravitreal injections so you can more readily observe the needle and its location within the vitreous cavity during the procedure.
4. Local anesthetic: 5 mg/mL (0.5%) proparacaine hydrochloride. Application of a local anesthetic to the eye prior to the procedure is recommended and will provide topical anesthetic during the procedure and shortly after.
5. Gel-based ocular lubricant (*see* **Note 2**): 3 mg/g hypromellose and 2.2 mg/kg carbomer 980. The use of an ocular lubricant allows for moisturization throughout the procedure, as well as magnification of the eye when performing the intravitreal injection.

6. Antibiotic solution: 5 mg/mL (0.5%) chloramphenicol, in either eye drop or ointment form. We apply an antibiotic to the mouse's eye immediately after the intravitreal injections are completed to help prevent superficial bacterial infections in the days following the procedure.

2.2 Tools Required for the Procedure

1. Stereoscopic microscope (*see* **Note 3**).
2. Heat mat.
3. A pair of straight and smooth tipped forceps (*see* **Note 4**).
4. A pair of straight forceps with teeth (*see* **Note 4**).
5. Beveled, sharp needle, approximately 29-gauge (*see* **Note 5**).
6. 10 μL syringe.
7. Blunt needle, approximately 35-gauge (placed in the 10 μL syringe; *see* **Note 6**).
8. Ophthalmic PVA spears.

2.3 Optional Tools Required for Unassisted Intravitreal Injections

1. Microinjection syringe pump and appropriate controller.
2. Foot switch.
3. Tubing.
4. Needle holder.

3 Methods

3.1 Preparation Before the Start of Procedure in Adult Mice

1. Adequately anesthetize your mouse.
2. While the anesthesia is still taking effect, apply the 0.5% tropicamide solution to both eyes of the mouse.
3. Once the bipedal reflex is absent, position your mouse underneath the dissecting microscope, on top of a heat mat. Position the mouse horizontally across the heat mat, with its head facing your nondominant hand and the tail facing your dominant hand. The mouse can either lay flat on its stomach or slightly on its side for easier access to the eye that you are injecting.
4. Confirm the pupils are dilated from application of the 0.5% tropicamide.
5. Once the pupils are dilated, use an ophthalmic PVA spear to remove the drops.
6. Apply a drop of 0.5% proparacaine hydrochloride to each eye, and wait approximately 5 min for the drug to take effect.
7. During these 5 min, set up your surgical instruments and syringes as per Fig. 1. We place the straight forceps with teeth next to the microscope, on the side of our nondominant hand, while the straight and smooth tipped forceps, gel-based ocular

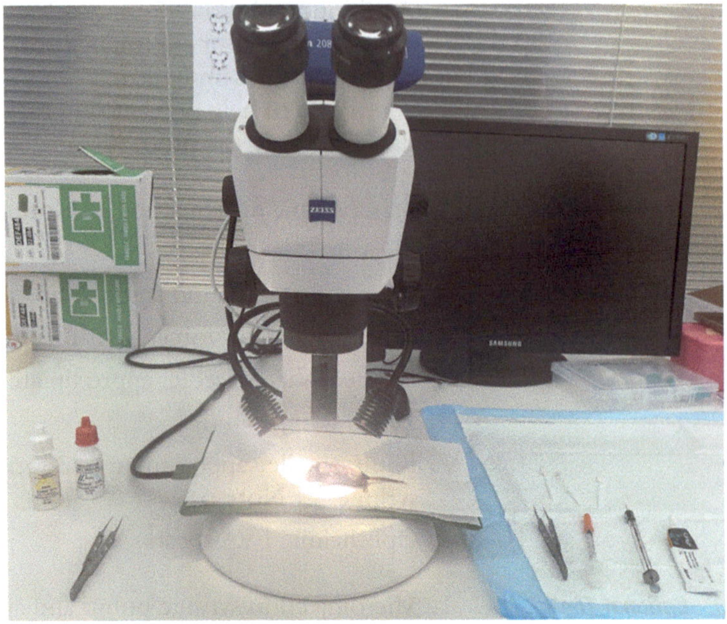

Fig. 1 Suggested setup for intravitreal injections. On the side of your non-dominant hand, place the straight forceps with teeth. On the side of your dominant hand, place the straight and smooth tipped forceps, gel-based ocular lubricant, ophthalmic PVA spears, beveled sharp needle, and the blunt needle (inserted into a syringe filled with your treatment)

lubricant, beveled sharp needle, and blunt needle (inserted into the 10 μL syringe, filled with the treatment you are providing) will be on the other side of the microscope.

3.2 Alternative Preparation Before the Start of Procedure in Preweaned Mice with Closed Eyes

Mouse pups do not naturally open their eyes until approximately 12–14 days after birth. Injections at this age, or before, are sometimes required when working with mouse lines that undergo rapid retinal degeneration. Below we describe how to open the mouse eye to prepare for an intravitreal injection.

1. Adequately anesthetize your mouse.
2. While waiting for the anesthesia to take effect, apply a drop of 0.5% proparacaine hydrochloride to each eye, and wait approximately 5 min for the drug to take effect.
3. Once the bipedal reflex is absent, use a beveled, sharp 29-gauge needle and run the needle across the natural line that is found where the top and bottom eyelids are fused together (Fig. 2a and b). Ensure you are opening the eye across the whole line but not exceeding this area unnecessarily (Fig. 2a–d; *see* **Note** 7).

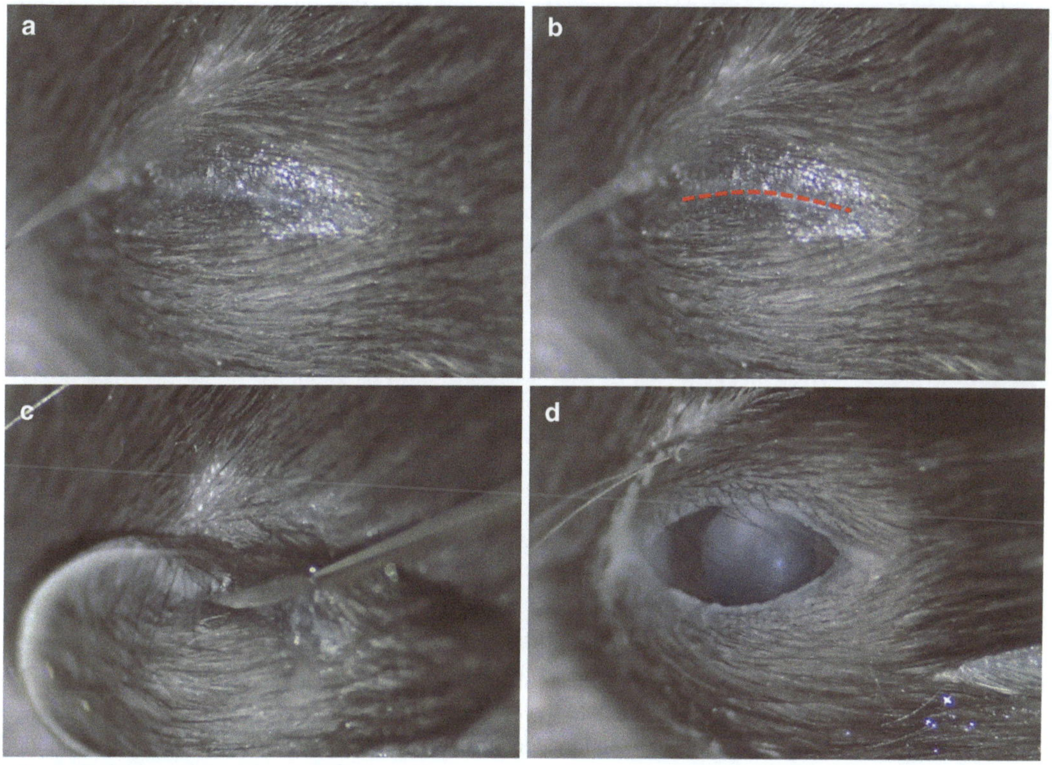

Fig. 2 (**a**) The sealed eye of a P12 mouse pup, with a natural groove that can be manually opened by the operator. (**b**) The sealed eye of a P12 mouse pup, with a natural groove that can be manually opened by the operator, with a red dotted line showing where the operator should open the eye. (**c**) Image depicting a pup's closed eye with the topical anesthesia 0.5% proparacaine hydrochloride solution and a 29-gauge beveled, sharp needle which the operator runs along the natural line to break the eyelid fusion. (**d**) The P12 pup eye fully opened after the eyelid fusion has been disrupted using a beveled, sharp needle. Operators can continue the intravitreal injection protocol as normal once the eye is open to this extent

4. Once the two eyelids have been separated, if the 0.5% proparacaine hydrochloride solution has been rubbed away, add an extra droplet (*see* **Note 8**).

3.3 Proptosis of the Eye

To access the side of the eye for injection, you will need to slightly proptose the eye.

1. With your dominant hand, use your straight and smooth tipped forceps with the tips of the forceps touching each other and place them on the lateral canthus of the eye (the junction between the two eyelids, closest to the mouse's temple, demonstrated in Fig. 3a). Push down gently and the eye should proptose easily (*see* **Note 9**).
2. Once you have proptosed the eye, keep your dominant hand pressing down gently to maintain the proptosed position of the eye. Using your nondominant hand, use your straight forceps

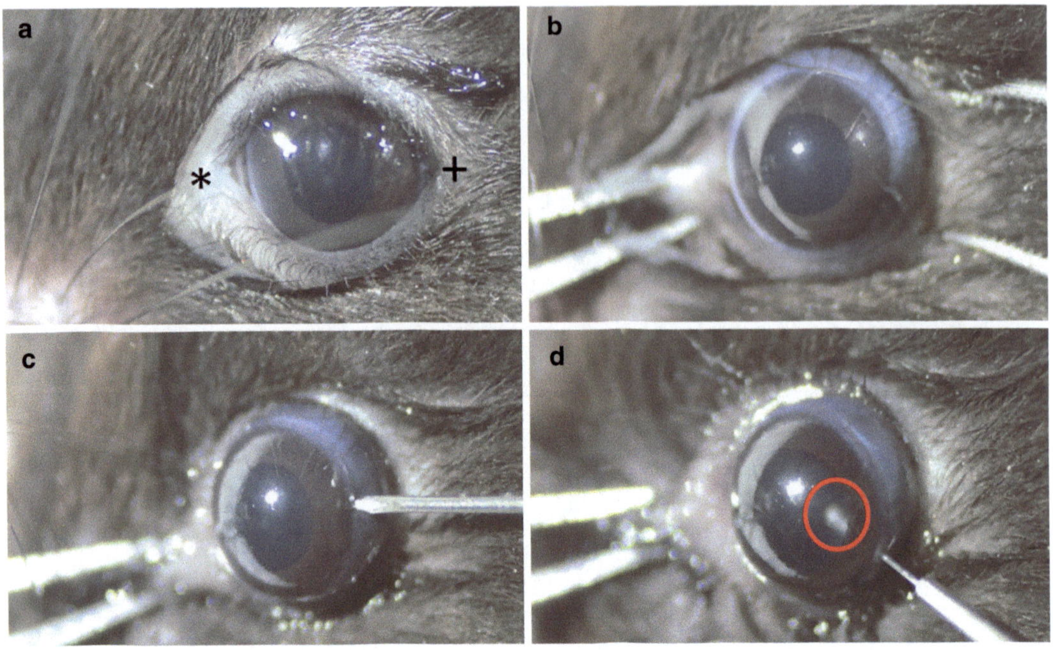

Fig. 3 An overview of the steps involving proptosis of the eye and administration of the treatment via the vitreous cavity. (**a**) Image of the mouse eye before the procedure demonstrating the medial canthus (denoted by a * symbol) and the lateral canthus (denoted by a + symbol). (**b**) To proptose the eye, use a straight and smooth tipped pair of forceps in your dominant hand to apply pressure to the lateral canthus, which allows the eye to pop out. Once the eye is proptosed, use your forceps with teeth to hold the tissue around the medial canthus with your nondominant hand, holding firmly and close to the eye so the eye remains proptosed for the next steps. (**c**) While still holding the medial canthus with your nondominant hand, use your dominant hand to make a small cut on the temporal pars plana with a beveled sharp 29-gauge needle. *NB: In this step, you will usually apply an ocular lubricant to help moisturize the eye and magnify it. For demonstration purposes, we did not apply it so the image can be seen better. (**d**) While still holding the medial canthus with your nondominant hand, use a blunt 35-gauge needle and insert it into the hole just made with the 29-gauge beveled, sharp needle using your dominant hand. Once inside the eye, you will enter the vitreous cavity where you want to position the needle at the back of the vitreous cavity while avoiding touching the lens. As the mouse's pupil is dilated, you can often visualize where the needle is located (red circle). *NB: In this step, you will usually apply an ocular lubricant to help moisturize the eye and magnify it. For demonstration purposes, we did not apply it, so the demonstration image was clearer

with teeth to hold the skin around the medial canthus (the junction between the two eyelids, closest to the mouse's nose, demonstrated in Fig. 3a and b). Ensure you are holding quite close to the eye (Fig. 3b).

3. Stop pushing down on the lateral canthus with your straight and smooth tipped forceps. The eye should now be held in its proptosed position by your straight forceps with teeth. You will need to ensure your nondominant hands keep a grip on the medial canthus for the remainder of the procedure. As such, all of the following steps will be performed using your dominant hand.

4. Add a drop of ocular lubricant to the eye.
5. Using your beveled sharp needle, puncture a hole at the temporal pars plana (border between the sclera and the cornea), being sure the hole is not too close to the iris, the retro-orbital vein, nor the sclera of the eye (target area is demonstrated in Fig. 3c). Do not put the needle all the way in; it should only be put deep enough to produce a small pilot hole.
6. Use your 35-gauge blunt needle, which you will already have connected to a syringe that contains your therapeutic. Insert the blunt needle into the pilot hole. Through the hole, you will enter the vitreous cavity of the eye, where you want to place the needle to the side of the lens and further back in the vitreous. As your mouse's pupil is dilated, you can often visualize where the needle is in the eye (demonstrated in Fig. 3d).
7. Plunge the needle (*see* **Note 10**), and keep the needle inside of the eye for at least a few seconds to help reduce reflux of the drug out of the eye.
8. Remove the needle from the eye, being careful not to touch the lens.
9. Using your straight and smooth tipped forceps, place the mouse eye fully back into the socket.
10. Remove the ocular lubricant from the eye using an ophthalmic PVA spear.
11. Repeat this process on the other eye if you are injecting both eyes.
12. After the procedure is completed, apply antibacterial ointment or drops to each eye that has been injected.
13. Reverse the mouse's anesthesia.

4 Notes

1. We recommend using injectable anesthesia when performing an intravitreal injection because a nosepiece providing a continuous gas may make it harder to complete this procedure as it restricts ideal hand positioning.
2. We highly recommend using gel-based rather than a liquid-based lubricant in this procedure. Gel-based lubricants are advantageous as they both magnify the eye and help reduce the spread of any potential bleeding that can occur from accidentally nicking the retro-orbital vein. We find when the retro-orbital vein has been nicked and liquid lubricant is being used, the blood completely obscures the visual field making the procedure more challenging.

3. Ensure the stereoscopic microscope that you use has a total magnification that can be adjusted. Our microscope goes from 8× to 40×, with most users conducting intravitreal injections at 10× or 20×.
4. The decision of which forceps to use for this procedure is personal; however, we recommend one pair of straight and smooth tipped forceps and another straight pair of forceps that has teeth. The straight and smooth tipped forceps will be appropriate to use when initially proptosing the eye, while the straight forceps with teeth help to hold the eye in a proptosed position for the whole procedure as it allows you to keep a good grip, consequently preventing the eye slipping back into the socket.
5. We recommend using a 29-gauge beveled sharp needle, as the hole made with this gauge is not large enough to damage the eye but is big enough for the blunt needle to fit through when you are performing the intravitreal injection. You can use just the needle, or alternatively insulin syringes can be used with the attached 29-gauge needles. This allows for a nice grip on the syringe shaft to have more control, compared to using a needle without the syringe attachment.
6. Ensure you use a blunt needle that has a relatively large gauge (using a larger gauge than the sharp needle that you used to make the initial hole is essential). We use a 35-gauge needle, which fits quite well into the pilot hole that is made with the 29-gauge beveled, sharp needle in the previous step.
7. In mice that are around 12–14 days old, the fusion between the eyelids will have already started to break down so you will not have to do too much work to open the eyes. However, in mice that are younger, you may need to do this process a few times as the eyelids are more sealed. Continue to only run the needle across rather than digging downward too much, so as not to damage the anterior chamber of the eye. It is also very important to make sure you follow the natural eyelid groove, as opening the eyes too wide or not enough will make eye proptosis in the next step challenging.
8. The use of pupil dilating solutions is not required when performing intravitreal injections on pups that had their eyes naturally closed, as their pupils will already be dilated.
9. If the eye will not move out of the socket, then adjust where your forceps are located and try again. Avoid placing too much downward pressure as only light pressure is required, meaning if the eye will not proptose, there may be an issue with how the forceps are positioned. If you are performing this surgery on young mice where you had to manually open their eyes, then check the eye is opened fully, following along the natural lines of the fused eyelids.

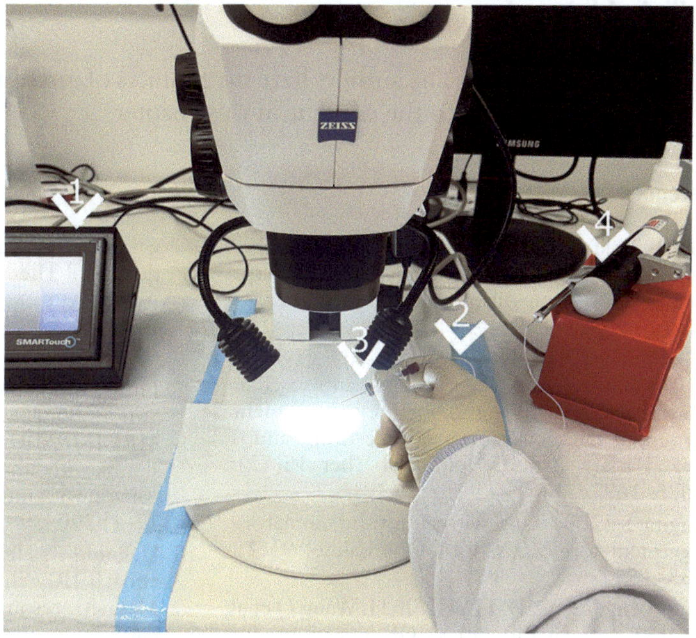

Fig. 4 Suggested setup with our preferred pump system for unassisted intravitreal injection. If using a pump system that allows for unassisted intravitreal injection, place the pump system and the needle with components attached on the side of your dominant hand before you start the procedure. The controller is placed away on the side of your nondominant hand as the foot pedal is used to control the delivery. Labeled as number 1 is our preferred controller for the pump (MICRO2T SMARTouch™, World Precision Instruments, Florida, USA), number 2 is the tubing we use (Silflex Tubing, SILFLEX-2, World Precision Instruments), number 3 is the needle holder (Nanofil Injection Holder, World Precision Instruments), and number 4 is the syringe attached to the pump (UMP3 syringe pump, World Precision Instruments)

10. When you plunge the 10 μL syringe, you will either need an assistant to plunge the needle for you, or you can perform the procedure unassisted with an automatic foot pump system (Fig. 4). You will use a needle that will be placed into the needle holder and connected to a tube and syringe barrel. When selecting the tubing, it is important to consider the size and diameter of the tube as this will affect the amount of treatment you will need to prepare. Prefill all components with the treatment you will be administering and attach the syringe to the pump system. We administer treatment in preweaned mice at a rate of 100 nL/s, while adult mice have treatment administered at a rate of 250 nL/s. Once the needle is in the vitreous cavity, the foot pedal is used to administer the drug unassisted.

Conflict of Interest

The authors have no conflicts of interest to declare that are relevant to the content of this chapter.

References

1. Ochakovski GA, Bartz-Schmidt KU, Fischer MD (2017) Retinal gene therapy: surgical vector delivery in the translation to clinical trials. Front Neurosci 11:174
2. Chatziralli I, Nicholson L, Sivaprasad S, Hykin P (2015) Intravitreal steroid and anti-vascular endothelial growth agents for the management of retinal vein occlusion: evidence from randomized trials. Expert Opin Biol Ther 15(12): 1685–1697
3. Sugita S (2007) Intravitreal anti-inflammatory treatment for uveitis. Br J Ophthalmol 91(2): 135–136
4. Ba J, Peng RS, Xu D, Li YH, Shi H, Wang Q et al (2015) Intravitreal anti-VEGF injections for treating wet age-related macular degeneration: a systematic review and meta-analysis. Drug Des Devel Ther 9:5397–5405
5. Maclachlan TK, Lukason M, Collins M, Munger R, Isenberger E, Rogers C et al (2011) Preclinical safety evaluation of AAV2-sFLT01- a gene therapy for age-related macular degeneration. Mol Ther 19(2):326–334
6. Pardue MT, Allen RS (2018) Neuroprotective strategies for retinal disease. Prog Retin Eye Res 65:50–76
7. Lin TH, Chiang CW, Trinkaus K, Spees WM, Sun P, Song SK (2014) Manganese-enhanced MRI (MEMRI) via topical loading of Mn(2+) significantly impairs mouse visual acuity: a comparison with intravitreal injection. NMR Biomed 27(4):390–398
8. Haenold R, Herrmann K-H, Schmidt S, Reichenbach JR, Schmidt K-F, Löwel S et al (2012) Magnetic resonance imaging of the mouse visual pathway for in vivo studies of degeneration and regeneration in the CNS. NeuroImage 59(1): 363–376
9. Morlet N, Young SH (1993) Prevention of intraocular pressure rise following intravitreal injection. Br J Ophthalmol 77(9):572

Chapter 12

Behavioral Assessment of Vision in Dogs Using a Four-Alternative, Choice-Based Testing Device

Freya M. Mowat and Simon M. Petersen-Jones

Abstract

Dogs are important large animal models of human retinal disorders. Visual navigation ability is an important outcome measure with relevance to real-world conditions. Here we describe methods to conduct behavioral assessment of dog visual navigation ability in a variety of luminance conditions. This method is suitable as a therapeutic trial endpoint for gene therapy studies in dogs.

Key words Canine, Visual assessment, Behavior, Maze, Tunnel, Translational model

1 Introduction

Dogs are well established as important translational laboratory-based models of human inherited retinal disorders [1]. In addition, outbred companion dogs are emerging as naturally occurring models of human visual aging [2, 3]. Dogs are a useful comparative model species; like humans, the dog retina is rod-dominated and contains a cone-rich area centralis with similarities to the human foveomacular retinal region [4, 5]. Dog color vision is dichromatic [6], and both brightness discrimination and visual acuity are lower than humans, so when using visual or cognitive tasks in dogs with a visual component, consideration should be made to these differences [7]. There are benefits to behavioral testing as a clinical outcome measure to complement electrophysiological techniques, tests of visual acuity, or object recognition tasks. Behavioral visual navigation tasks provide information on an integrated skill that has relevance to real-world conditions. Standardized visual navigation tasks have also been developed as human therapeutic clinical trial endpoints for the same reasons [8]. These tasks can be implemented in nonverbal animals (including dogs), can be performed awake without sedation or anesthesia, and typically require no prior training to complete.

Livia S. Carvalho (ed.), *Retinal Gene Therapy: Methods and Protocols*, Methods in Molecular Biology, vol. 3023,
https://doi.org/10.1007/978-1-0716-5198-8_12,
© The Author(s), under exclusive license to Springer Science+Business Media, LLC, part of Springer Nature 2026

Two such methods have been described for use in dogs, both of which have been validated as therapeutic trial endpoints for retinal gene augmentation therapy. The first is a standardized obstacle avoidance task [9], whereby dogs must navigate a series of vertical panels to successfully exit a 3.6-meter-long maze device. Dogs are not trained on the device but are typically conditioned to the testing environment and trained to respond to vocal, toy, and/or treat-based rewards used as motivation during testing. Testing can be completed in varying luminance, and this method is a sensitive outcome measure for detection of heritable forms of blindness in dogs [9, 10] and in retinal gene augmentation clinical studies [11]. The second is a four-alternative, choice-based visual testing system, which leverages a dog's natural motivation to exit a confined space. Original descriptions of the utility of the device to detect dogs with heritable retinal disorders demonstrated good sensitivity [12] and reproducibility [13] with no prior training or acclimation. Subsequently, the device has been used in numerous studies as a clinical trial endpoint in retinal therapeutic trials in laboratory dogs [14, 15] and in outbred companion dogs [16]. Methods to implement this device in a laboratory setting are described. Dogs can be tested in a variety of lighting environments, and the open tunnel for exit navigation is randomized for each test trial. Multiple trials at each illuminance are recommended. Two outcome measures are recorded—the time taken to exit the device and the proportion of trials at each illuminance in which the correct (open) tunnel is selected first.

2 Materials

The materials to build the four-choice tunnel device must be gathered, and the device self-assembled as there are no commercial manufacturers at present. Figure 1 shows a schematic of the device and an image of the device in use.

2.1 Specialized Equipment

1. Central box: Each box side is 4 ft. long × 2 ft. 10″ tall and is typically constructed from 3/8" PVC panels. A 9″ hinged door is placed in the side of a corner of the box to facilitate dog entry. *See* **Note 1**.
2. Exit tunnels: The four tunnels are constructed of commercially available PVC coated vinyl, cuffed, wire-reinforced 24–26″ diameter round tunnels, and should be 3′ long. *See* **Note 2**. Secure tunnel caps should be constructed that obstruct light at the end of the tunnel completely, as this is the method for the dog to navigate out of the device. *See* **Note 3**.

3. Lighting: Each tunnel end is illuminated, irrespective of which tunnel is open during a trial. This allows for rapid trial repetition without extensive lighting adjustment. Materials required include four adjustable clamp attach lamps capable of accommodating up to 150 W bulbs that are attached to adjustable 6′ extendable tripods. Lights are situated 40″ from the floor above the tunnel end. Each lamp is controlled by a dimmer switch—self-made using on/off dimmer switches, an extension cord, junction box, and cap. *See* **Note 4**. Bulbs should be dimmable incandescent light bulbs emitting 25–100 W. *See* **Notes 5** and **6**. A light meter that measures illuminance is needed to calibrate lighting levels at the end of each tunnel. *See* **Note 7**.

2.2 Miscellaneous

1. Black, light-blocking fabric (theater cloth from fabric supplier) is used to manage light leaks as needed.
2. Stopwatch (no backlit display) for recording time to exit the device.
3. Electrical extension cord with multiple outlets and breaker to control power to lighting dimmer switches and light meter.
4. Measuring tape to match lamp height to 40″.
5. Data recording sheets and clipboard.
6. Photo-safe red headlamps, safelights, or flashlights for working in scotopic conditions if needed.
7. Disposable gloves, disinfectant cleaning spray, mop, and paper towel in case of animal elimination and at the end of testing. *See* **Note 8**.
8. A video camera for recording trials for post hoc review/analysis. *See* **Note 9**.

3 Methods

3.1 Testing Environment

1. The testing room should be at least 11 ft. at its shortest edge to allow sufficient space around the end of tunnels for dogs to navigate out confidently and safely.
2. Attention should be paid to eliminate sources of stray light in the room (e.g., around door frames and ceiling tiles), particularly if low luminance testing is to be performed. Use of flat back interior paint and dark-colored flooring is recommended to eliminate light scattering.
3. Flooring should be sanitizable.
4. If low lighting within scotopic range is to be completed, animals should be acclimated to the lighting condition for 20 min, and dark adaptation time should be standardized for each experiment.

3.2 Setup

1. Set up the four-choice apparatus so that a tunnel extends fully from each wall of the apparatus and there is enough room at the end for a dog to exit the tunnel.
2. Place a lamp on a tripod beside the end of each tunnel so that the light bulb is 40″ above the floor and is shining down on the exit of the tunnel.
3. The tunnels are designated 1–4, starting to the right of the apparatus door and going clockwise. Figure 1a shows a schematic drawing of the system as viewed from above. Figure 1b shows a fully functional setup being used.
4. With room lights illuminated, open the box door and evaluate inside the box for presence of light leaks.
5. Use black fabric or similar to block stray light.

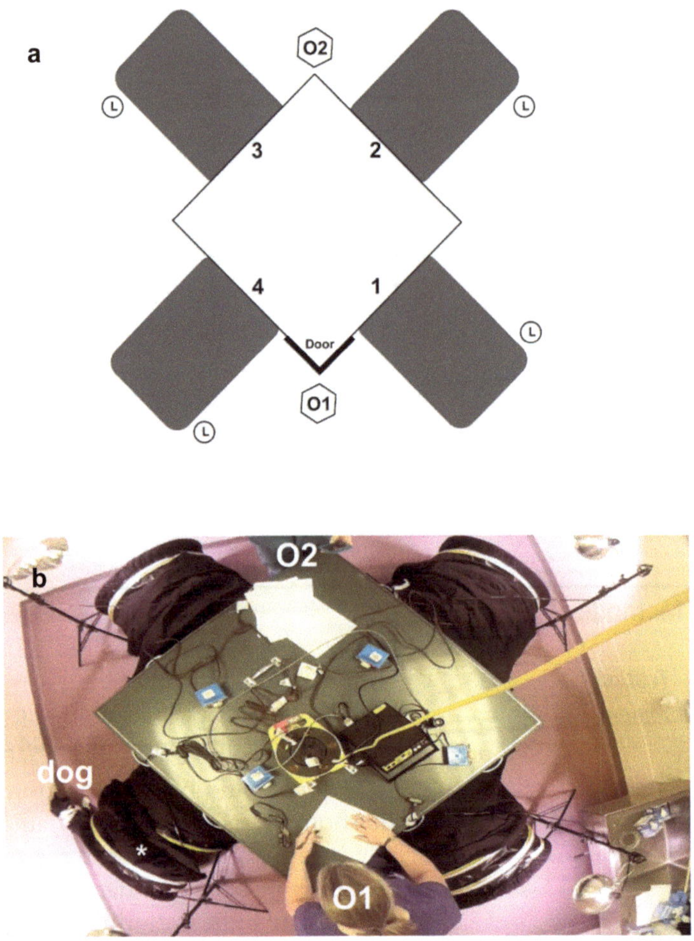

Fig. 1 (**a**) Schematic of testing setup, to scale. A 4 ft. box with 3 ft. long tunnels, 24 inches wide, enclosed in 11 ft. square room. (**b**) Wide-angle picture of built device in use. Dog (marked) is exiting the open tunnel 4 (marked with asterisk). L, lamp tripod; 01, observer 1; 02, observer 2

6. Using the light meter, measure the light intensity at the floor level in front of each tunnel exit (4) and adjust with the dimmer so that each light is at the desired setting for that series of trials.
7. Set each light at the approximate level first, and then repeat the measurements to make final adjustments for each light and record the measurements on the data form.
8. The light intensity order can be randomized ahead of time, to limit the effect of habituation as a potential confounding factor in analysis.

3.3 Testing

1. Open tunnel order and light intensity trial set order is randomized ahead of time.
2. Luminances selected are at the discretion of the investigator, depending on the hypothesis.
3. Once the dog is adapted to the lighting condition, begin the trials.
4. A trial is conducted by covering the ends of three designated tunnels with the fabric covers and then placing the dog into the central square box using the door.
5. Dog collars and leashes should be removed.
6. Lights are on at the pre-set intensity by all four tunnel exits.
7. Use a stopwatch to measure the length of time from when the dog is released into the box until the dog's nose crosses the exit of the open tunnel.
8. Also note which tunnel the dog enters first by listening and watching for movement (two observers are useful to monitor two tunnels each).
9. The maximum trial length is 60 s (a longer trial length may be adopted for companion dogs; *see* [16]). *See* **Note 10**.
10. Seven to eight consecutive trials are performed at each light intensity, with one tunnel end open for each trial in a random order of presentation.

4 Notes

1. The box can be constructed as completely enclosed, or with an open floor to facilitate easy floor cleaning. Each side panel can interconnect with adjacent panels permanently, or using a temporary attachment system (e.g., hinges) to allow for assembly and disassembly. The tunnel holes are cut slightly smaller than the tunnel diameter cut out of each side, 1 inch from the lower portion of each panel. Angle flanges could be used to attach tunnels to the box if desired, or the tunnel end can simply be inserted into the hole. Attention must be paid to potential light

leaks around the tunnel attachments to the box, and at the lower edge of the box, light-blocking fabric or similar material should be used to eliminate stray light, so that the box is as lightproof as possible.

2. Sturdy, dark, opaque plastic sanitizable material is recommended. This size is standard for commercially available dog agility tunnels and for ventilation ducts, and customizing is recommended to seal in the wire reinforced hoops at the terminal end.
3. If dogs can perceive weaknesses to the exits to tunnels, they may force their way through obstructed exits. Ensure that tunnel caps are secure and opaque. For lightproof tunnel ends, we fabricated elasticated tunnel ends from black opaque theater cloth that fit snugly over the end of closed tunnels and can be quickly removed.
4. Instructions for homemade dimmer switches can be found online (e.g., https://www.youtube.com/watch?v=_e9Rk7_B8GY).
5. Bulbs recommended include 25, 40, 60, and 100 watts to achieve a range of lighting intensities.
6. Safety considerations: Consult with an electrician when constructing electrical devices. Unplug all lights and the light meter when they are not in use. Light bulbs get hot! Use a towel or cloth to prevent burns when changing bulbs. A ceiling-mounted electrical supply can help to minimize trip hazards in the room for dogs and handlers.
7. We utilize the ILT1700 (now updated to the ILT5000; International Light Technologies, Peabody, MA, USA) light meter with SED033 probe and Barrel R #512, Diffuser W #13986, and Filter Y #29194.
8. Pheromone and odor elimination between individuals is important within the device and tunnels, as dogs will explore the environment with their olfaction which may skew trial outcomes early in testing. Strict disinfection and cleaning protocols are recommended.
9. It is preferable to select a digital camera with a wide recording angle that is ceiling mounted, low lighting suitable/infrared, and is capable of being remotely controlled. *See* Fig. 1b for an image taken from a suitably positioned camera.
10. Make notes of any behavior which may affect the response such as scratching at the tunnels, lying down, or eliminating. No rewards or external cues are used during the trial. After the trial is completed, the dog may be praised or petted, but no food treats are used.

Conflict of Interest

The authors have no conflicts of interest to declare that are relevant to the content of this chapter.

References

1. Winkler PA, Occelli LM, Petersen-Jones SM (2020) Large animal models of inherited retinal degenerations: a review. Cells 9(4). https://doi.org/10.3390/cells9040882
2. Salzman MM, Merten N, Panek WK, Fefer G, Mondino A, Westermeyer HD, Gruen ME, Olby NJ, Mowat FM (2023) Age-associated changes in electroretinography measures in companion dogs. Doc Ophthalmol. https://doi.org/10.1007/s10633-023-09938-7
3. Rogers CM, Salzman MM, Li Z, Merten N, Russell LJ, Lillesand HK, Mowat FM (2023) Subjective vision assessment in companion dogs using dogVLQ demonstrates age-associated visual dysfunction. Front Vet Sci 10. https://doi.org/10.3389/fvets.2023.1244518
4. Mowat FM, Petersen-Jones SM, Williamson H, Williams DL, Luthert PJ, Ali RR, Bainbridge JW (2008) Topographical characterization of cone photoreceptors and the area centralis of the canine retina. Mol Vis 14:2518–2527
5. Beltran WA, Cideciyan AV, Guziewicz KE, Iwabe S, Swider M, Scott EM, Savina SV, Ruthel G, Stefano F, Zhang L, Zorger R, Sumaroka A, Jacobson SG, Aguirre GD (2014) Canine retina has a primate fovea-like bouquet of cone photoreceptors which is affected by inherited macular degenerations. PLoS One 9(3):e90390. https://doi.org/10.1371/journal.pone.0090390
6. Neitz J, Geist T, Jacobs GH (1989) Color vision in the dog. Vis Neurosci 3(2):119–125. https://doi.org/10.1017/s0952523800004430
7. Pongrácz P, Ujvári V, Faragó T, Miklósi Á, Péter A (2017) Do you see what I see? The difference between dog and human visual perception may affect the outcome of experiments. Behav Process 140:53–60. https://doi.org/10.1016/j.beproc.2017.04.002
8. Kumaran N, Ali RR, Tyler NA, Bainbridge JWB, Michaelides M, Rubin GS (2020) Validation of a vision-guided mobility assessment for RPE65-associated retinal dystrophy. Transl Vis Sci Technol 9(10):5. https://doi.org/10.1167/tvst.9.10.5
9. Garcia MM, Ying GS, Cocores CA, Tanaka JC, Komáromy AM (2010) Evaluation of a behavioral method for objective vision testing and identification of achromatopsia in dogs. Am J Vet Res 71(1):97–102. https://doi.org/10.2460/ajvr.71.1.97
10. Iwabe S, Ying GS, Aguirre GD, Beltran WA (2016) Assessment of visual function and retinal structure following acute light exposure in the light sensitive T4R rhodopsin mutant dog. Exp Eye Res 146:341–353. https://doi.org/10.1016/j.exer.2016.04.006
11. Miyadera K, Santana E, Roszak K, Iffrig S, Visel M, Iwabe S, Boyd RF, Bartoe JT, Sato Y, Gray A, Ripolles-Garcia A, Dufour VL, Byrne LC, Flannery JG, Beltran WA, Aguirre GD (2022) Targeting ON-bipolar cells by AAV gene therapy stably reverses LRIT3-congenital stationary night blindness. Proc Natl Acad Sci USA 119(13): e2117038119. https://doi.org/10.1073/pnas.2117038119
12. Gearhart PM, Gearhart CC, Petersen-Jones SM (2008) A novel method for objective vision testing in canine models of inherited retinal disease. Invest Ophthalmol Vis Sci 49(8): 3568–3576. https://doi.org/10.1167/iovs.07-0625
13. Annear MJ, Gornik KR, Venturi FL, Hauptman JG, Bartoe JT, Petersen-Jones SM (2013) Reproducibility of an objective four-choice canine vision testing technique that assesses vision at differing light intensities. Vet Ophthalmol 16(5):324–328. https://doi.org/10.1111/j.1463-5224.2012.01076.x
14. Gearhart PM, Gearhart C, Thompson DA, Petersen-Jones SM (2010) Improvement of visual performance with intravitreal administration of 9-cis-retinal in Rpe65-mutant dogs. Arch Ophthalmol 128(11):1442–1448. https://doi.org/10.1001/archophthalmol.2010.210
15. Occelli LM, Zobel L, Stoddard J, Wagner J, Pasmanter N, Querubin J, Renner LM, Reynaga R, Winkler PA, Sun K, Marinho L, O'Riordan CR, Frederick A, Lauer A, Tsang SH, Hauswirth WW, McGill TJ, Neuringer M, Michalakis S, Petersen-Jones

SM (2023) Development of a translatable gene augmentation therapy for CNGB1-retinitis pigmentosa. Mol Ther 31(7):2028–2041. https://doi.org/10.1016/j.ymthe.2023.04.005

16. Young WM, Oh A, Williams JG, Foster ML, Miller WW, Lunn KF, Mowat FM (2018) Clinical therapeutic efficacy of mycophenolate mofetil in the treatment of SARDS in dogs-a prospective open-label pilot study. Vet Ophthalmol 21(6):565–576. https://doi.org/10.1111/vop.12545

Chapter 13

Behavioral Vision Testing Using an Automated Observational Tracking System (VisioBox) to Obtain a Measure of Visual Acuity in Larval Zebrafish

Emily L. Allan and Jessica K. Mountford

Abstract

The optokinetic response (OKR) within zebrafish larvae consists of automatic stereotypical eye movements in response to a visual stimulus. This response is characterized by a tracking (slow phase) followed by a rapid (fast phase) saccade in the opposite direction to the stimulus. Studying OKR behavior provides valuable information on the visual function of genetically modified animals. Here we describe a method for assessing visual function and color vision testing in zebrafish larvae at 5 days postfertilization.

Key words Optokinetic response, Zebrafish, Visual acuity, Color vision, Ocular disease modeling

1 Introduction

The optokinetic response is a compensatory reflex that supports visual image stabilization, characterized by phases of stimulus tracking or slow-phase velocity (SPV), followed by a fast-phase velocity (FPV) eye movement whereby the eye rapidly returns to a resting position. During this FPV, the position of the eye is returned to resting via a saccade in the opposite direction to the stimulus. Optokinetic response testing of larval zebrafish has become an attractive way of gaining measures of visual acuity and color vision within a biological cohort expressing abnormal ocular defects due to genetic aberrations. The VisioBox (ViewPoint, France) is an automated observational tracking system which allows researchers to test the optokinetic response of larval zebrafish rapidly and accurately. The customizable testing protocol allows researchers to tailor the test to the specific needs of the larvae, optimizing the accuracy and applicability of results. A lack of standardized reporting for optokinetic response research in larval zebrafish has led to gaps in knowledge and research with an inability to be replicated.

Livia S. Carvalho (ed.), *Retinal Gene Therapy: Methods and Protocols*, Methods in Molecular Biology, vol. 3023,
https://doi.org/10.1007/978-1-0716-5198-8_13,
© The Author(s), under exclusive license to Springer Science+Business Media, LLC, part of Springer Nature 2026

Here, we demonstrate a recommended testing and reporting protocol using AB strain wild-type larval zebrafish at 5 days postfertilization (dpf).

1.1 Zebrafish as a Model for Ocular Research

The visual system of zebrafish is significantly like that of humans, with shared functional and developmental vision mechanisms [1]. Zebrafish tetrachromatic color vision is comparable to that of human trichromatic vision, with the addition of UV photoreceptor cone visual opsin gene (*opn1sw1*). The markedly large zebrafish eyes undergo rapid development, commencing at 24 h postfertilization (hpf), and become fully functional by 5 dpf, rendering them an advantageous model for studying ocular disease [2]. In addition, zebrafish have external, transparent embryogenesis vital for developmental studies; are space and cost-effective to maintain; and produce abundant egg clutches (up to 200 eggs per spawning female), facilitating rapid screening of candidate genes within a controlled biological cohort [3].

2 Materials

Prepare all solutions using ultrapure H_2O and molecular grade reagents to ensure sterility for in vivo use. All reagents are to be prepared at room temperature unless otherwise stated below.

2.1 E3 Medium (60×)

1. Dissolve 34.8 g NaCl; 1.6 g KCl; 5.8 g $CaCl_2.2H_2O$; and 7.78 g $MgCl_2.6H_2O$ into 2 L of ddiH_2O. Adjust pH to 7.2 with NaOH and autoclave.
2. For working stocks, dilute to 1× by adding 16.5 mL of the 60× stock into ddiH_2O.
3. If using methylene blue for antifungal and antimicrobial properties and maintaining a sterile environment for the developing embryos and larvae, add 100 μL of 1% methylene blue to the 1 L of 1× E3 solution, 0.00001% final concentration (*see* **Note 1**).

2.2 Danieau's Solution (30×)

1. Dissolve 101.7 g NaCl; 1.56 g KCl; 2.96 g $MgSO_4{\bullet}7H_2O$; 4.25 g Ca $(NO_3)_2$; and 35.75 g HEPES into 1 L of ddiH_2O. Adjust pH to 7.6.
2. Dilute to 1× working stock by adding 33.3 mL of 30× stock into 1 L ddiH_2O for experimental use.
3. Store 30× stock solution at 4 °C.

2.3 System Water

1. For consistency, follow the Tecniplast system method for automatically dosing for conductivity, 800 μS +/− 100 μS (salts, Aquarium Systems Instant Ocean Aquarium Salt, France), pH 7.4 +/− 0.6 (sodium bicarbonate), and heating elements for the water temperature, with a set point 28.5 °C +/− 1.5 °C.

2. Test system water weekly using a water testing kit (API Liquid Test Kits, API) for general hardness (GH, 71.6–89.5 mg/L), carbonate hardness (KH, 17.9–35.8 mg/L), ammonia (<0.05 mg/L), nitrates (<100 mg/L), and nitrites (target 0 mg/L) and a colorimeter (DR 900, HACH) for pH, ammonia (<0.05 mg/L), nitrates (<100 mg/L), nitrites (target 0 mg/L), and oxygen (>6 mg/L).

2.4 3% Methylcellulose

1. Use 50 mL of either 1×E3 medium without methylene blue or 1× Danieau's solution and heat to 30 °C in a water bath.
2. Place an additional 50 mL of liquid (either 1× E3 medium without methylene blue or 1× Danieau's solution) on ice, cooling it to 1–4 °C.
3. Remove the warmed solution from the water bath and, while stirring, add 3 g of methylcellulose gradually [4].
4. Add the cooled solution to the methylcellulose mixture, using a magnetic stir plate to stir vigorously overnight at 4 °C in a cool room (*see* **Note 2**).
5. Once most of the methylcellulose has dissolved, store at 4 °C until needed.

2.5 Tricaine Methanesulfonate (MS-222)

Buffered MS-222 is to be prepared in a biological safety cabinet, ensuring the following PPE is worn, gloves, lab gown, and a respirator to prevent inhalation of particles.

1. Prepare stock solution of 8.4 g/L MS-222 by dissolving 1.68 g of MS-222 (Sigma-Aldrich) into 200 mL of reverse osmosis water.
2. Wrap 10 mL aliquots of 8.4 g/L stock solution in foil and store frozen prior to use.
3. To prepare the anesthetic bath, add 10 mL of thawed MS-222 stock solution (8.4 g/L) to 490 mL of system water to achieve final concentration of 168 mg/L MS-222.
4. Buffer the anesthetic bath using drops of 1 M sodium bicarbonate until the pH reaches between 7.0 and 7.5.
5. To anesthetize, place larvae in the MS-222 anesthetic solution for at least 5 min prior to experimental use.

3 Methods

3.1 Animal Use and Ethics Statement

For the purpose of this methods chapter, zebrafish (*Danio rerio*) of the AB strain were maintained and bred at the Western Australian Zebrafish Experimental Research Centre (WAZERC) at the University of Western Australia's Biomedical Research Facility according to the local animal standard operation procedures and guidelines. Adult AB strain zebrafish were maintained on a 14:10 h light/dark cycle at composition conductivity 800 μS +/− 100 μS (salts, Aquarium Systems Instant Ocean Aquarium Salt, France) and pH 7.4 +/− 0.6 (sodium bicarbonate) and heating elements for the water temperature, with a set point 28.5°C +/− 1.5°C (*see* **Note 3**).

1. To spawn sexually mature AB zebrafish >3 months postfertilization (mpf), separate the male and female (*see* **Note 4**) pairs using a clear plastic divider within the breeder box filled with system water the evening prior to spawning (*see* **Note 5**).
2. Remove dividers the following morning after lights on to facilitate spawning. Fertilized embryos filter through the grated inserts to separate from the adult zebrafish.
3. Collect embryos using a fine mesh sieve rinsed with E3 media to clean residue from spawning.
4. Once clean, transfer embryos at a density of 50 embryos per 90 mm Petri dish containing approximately 30 mL of 1× E3 media and place within an incubator kept at 28.5 °C on a 14:10 h light/dark cycle until reaching 5 dpf. Note that sex is not yet established during the larval stage.

3.2 Testing Visual Acuity Using VisioBox (ViewPoint)

To evaluate visual system performance in zebrafish larvae, the optokinetic response (OKR) is a reliable and sensitive measure of stereotypical eye movements. Precise assessment of visual function in zebrafish larvae necessitates a process that is smooth, controllable, and adaptable, enabling the detection of subtle visual defects. Measuring OKR requires comprehensive control over the zebrafish environment and behavioral conditions, with flexibility to adjust parameters such as color, light intensity, stimulus speed, and spatial frequency. Additionally, automated software capable of tracking visual acuity and contrast sensitivity through precise eye-detection algorithms is desirable for a comprehensive understanding of visual function.

To address these requirements, ViewPoint developed the VisioBox, a refined system optimized for zebrafish larval OKR. This platform features a flexible, high-contrast screen, a compact camera for enhanced resolution, and improved plateholder ergonomics to streamline experimental workflows. The design reflects extensive collaboration and is tailored to meet demands for accuracy, robustness, user-friendliness, and efficient data analysis. Additionally, the versatility of the system allows for conversion to optomotor response (OMR) testing in older zebrafish, extending its utility across developmental stages.

Methods for testing the automated observational tracking system (VisioBox, ViewPoint) should be conducted as per ViewPoint's standard operating procedures (*see* **Note 6**). The VisioBox

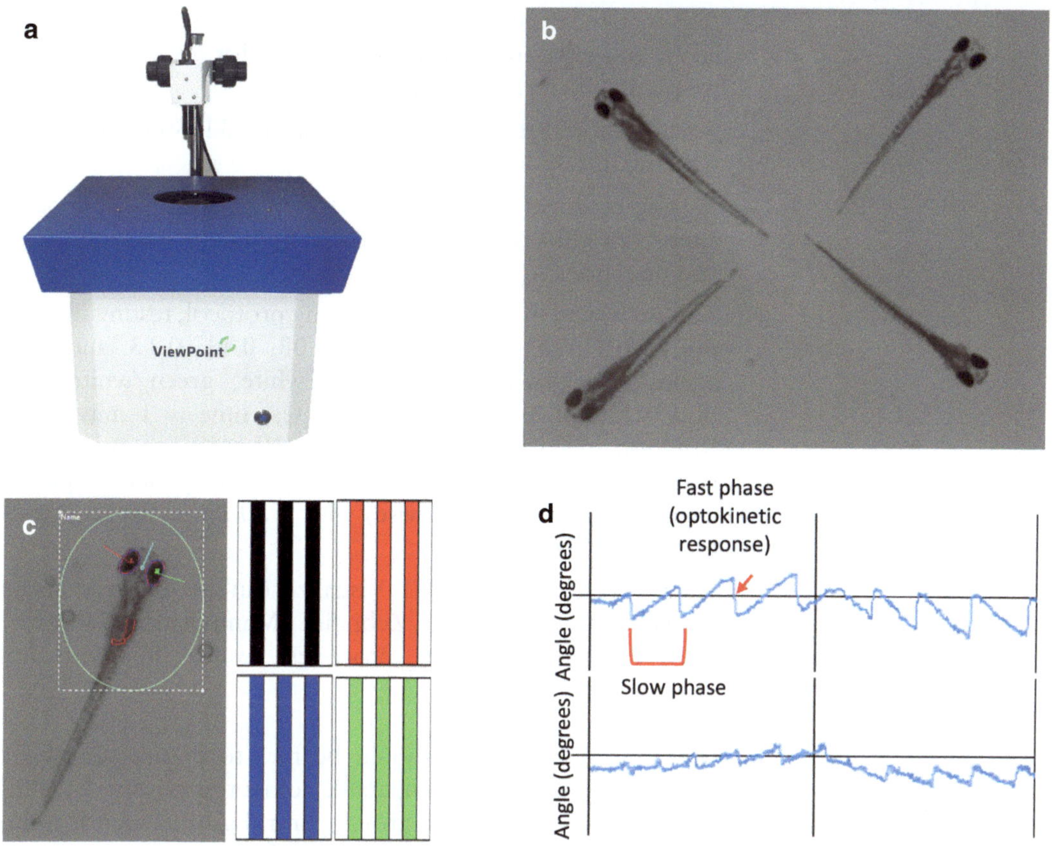

Fig. 1 Behavioral vision testing was conducted using an automated observational tracking system, VisioBox (ViewPoint) (**a**), to assess 5 dpf larvae for live optokinetic response and color vision. (**b**) Up to four larvae were immobilized using 3% methylcellulose within a 35 mm Petri dish placed on the stage. Vision was assessed using black/white, red/white, blue/white, and green/white color combinations (**c**), as well as spatial frequencies of 0.05, 0.08, 0.11, 0.13, and 0.16. (**d**) Example of both the fast- and slow-phase eye movement of the right eye representing optokinetic response of a wild type and morpholino knockdown larvae at 5 days postfertilization. The reduced range of motion in the knockdown larvae eye indicates a reduction in visual acuity

is comprised of three rectangular LED screens wrapped in a cylindrical shape around a stage (~50–58 mm from larvae eyes), with a circumference of 383 mm (Fig. 1). Screens display vertical stripes (with a brightness of 400–500 nits) rotating both clockwise and counterclockwise to stimulate the binocular vision of larval zebrafish.

1. To run a test, place four larvae at 5 dpf within a 35 mm Petri dish in a cross-like formation (facing larvae away from each other so that they are not within each other's visual field), immobilized in a 3% methylcellulose solution (*see* Subheading 2.4). Remove any bubbles within the methylcellulose using a Pasteur pipette to reduce refraction noise during testing.

2. Acclimatize larvae in darkness for 1 min prior to exposing the larvae to spatial frequencies of 0.05 cpd for 5 s in both clockwise and anticlockwise directions [5].
3. Set detection thresholds for each individual larva to enable eye tracking in response to stimuli.
4. Testing protocol can be customized using the inbuilt code to alter stripe color, orientation, spatial frequency, rotation speed, and direction.
5. Test each larva in triplicate per testing protocol, testing increasing spatial frequencies of 0.05, 0.08, 0.11, 0.13, and 0.16 against black/white stripes, red/white, green/white, and blue/white, changing every 20 s (test time = 1 min 41 s). Rotation speed remains constant at 10 rotations per minute (rpm), with directional changes at each spatial frequency change (starting as clockwise) with a 30-s rest between tests (*see* **Notes** 7 and **8**).
6. Data produced from PHI Visualize can be analyzed using Visual Interrogation of Zebrafish Manipulations software (VIZN), which rapidly processes results for statistical analysis [6].

3.3 Control Conditions

Certain control conditions should be met to ensure results are statistically significant and comparable across test groups. In addition to the main experimental group, control groups such as uninjected WT larvae and negative control vehicle- or sham- injected larvae should be utilized. Test groups should contain a minimal sample size of 12 larvae per group for statistical power analysis and come from the same biological cohort to reduce genetic variability. As above, larvae should be acclimatized to test conditions prior to testing and be allowed to rest between tests to reduce variability.

3.4 Statistical Analysis

Export eye angle data into excel or equivalent and sort results by spatial frequency. Using statistical analysis software, remove outliers and "impossible" data and use the average minimum and maximum eye angles to determine the range of motion of the eyes during testing. Sort data into temporal to nasal and nasal to temporal direction and combine with results from other test groups. This data can then be plotted and statistically analyzed in statistical analysis software such as GraphPad Prism. Bar graph best represents the nasal to temporal eye movements during testing (range of motion).

3.5 Data Reporting

To standardize larval OKR testing, below is a proposed a list of information that should be included in the presentation of results (Table 1). This will ensure researchers are producing transparent, replicable data.

Table 1
Proposed list of information to be included in results presentation

Information	Aspects to be reported	Example
Larval characteristics	1. Species and strain 2. Stage of development	1. Danio rerio, AB 2. 5 days postfertilization
Preparation of larva	1. Immobilization technique 2. Container and size 3. Position of larva	1. Methylcellulose, 3% 2. 35 mm petri dish 3. Facing outward in an X shape
OKR assay method	1. Manufacturer/model 2. Setup	1. VisioBox Optodrum 2. LED screens in a circular drum
Rotating drum	1. Drum material 2. Rotating method 3. Diameter	1. LED screens 2. Projected 3. 12 cm
Projection	1. Shape of screen 2. Projection technique	1. Circular 2. LED screens
LED arena	1. Shape of arena 2. Number of LED panels 3. Arrangement of panels	1. Circular 2. 3 3. Joined to create a cylinder
Stimulus	1. Stimulus pattern 2. Direction of stimulus rotation 3. Stimulus duration 4. Contrast 5. Eyes stimulated 6. Distance from larval eye 7. Light intensity	1. Gratings (vertical stripes) 2. Clockwise and anticlockwise 3. 1 min 41 s 4. Need to find out 5. Binocular 6. Need to find out 7. 400–500 nits
Experiment	1. Trials 2. Assay throughput 3. Sample size 4. Time of day	1. 3 per fish 2. $N = 3$ fish per group 3. Total larvae: 12 per group 4. 10–5 pm
Recording of eye movement	1. Camera	1. VisioBox inbuilt capture system
Analysis of eye rotation	1. Software 2. Outcome	1. VisioBox, excel, and prism 2. Temporal to nasal eye movement

4 Notes

1. It is important to note that adding methylene blue to E3 solution, even at low concentrations, can lead to changes that affect oxidative metabolism in the developing zebrafish larvae. Therefore, researchers interested in studies involving metabolic pathways need to be cautious when using methylene blue.
2. When preparing the 3% methylcellulose, use a large vessel such as a 200 mL wide-mouth glass tissue culture bottle with a large stir bar to help minimize bubble formation.

3. For the purpose of this chapter, all housing, rearing, and experimental procedures were conducted in line with the Australian National Health and Medical Research Council Code of Practice for the Care and Use of Animals in Research. Animal ethics approval for housing and breeding the adult zebrafish was obtained from the University of Western Australia (2021/ET000496) and OGTR/IBC approval of a NLRD via the University of Western Australia (2021/ET000446). Ensure the appropriate animal ethics and IBC approvals are sought in line with your institution prior to conducting animal research.
4. The most efficient method for spawning zebrafish is in pair-wise crosses, typically consisting of higher female-to-male ratios. Spawning efficiency decreases as the number of males increases.
5. Zebrafish rely upon several cues for spawning and reproduction, including light. The AB strain used in this example is sensitive to photoperiods, dawn in particular. Therefore, the "lights on" period mimics dawn in a laboratory setting, and spawning will typically commence.
6. Visual OKR testing is best conducted from 5 dpf onward as this is when the eye is fully functional and the larvae begin free feeding and using visual cues for hunting.
7. Power analysis needs to be conducted to ascertain the required number of larvae per treatment group needed to obtain statistical significance.
8. Aim to conduct the OKR testing between 9 am and 5 pm (kept consistent between experiments) on larvae from the same spawning batch to reduce potential variation in genetics and changes in developmental age due to the rapid development of the larvae.

Conflicts of Interest

The authors have no conflicts of interest to declare that are relevant to the content of this chapter.

References

1. Chinen A, Hamaoka T, Yamada Y, Kawamura S (2003) Gene duplication and spectral diversification of cone visual pigments of zebrafish. Genetics. [cited 2024 Aug 6];163(2): 663–675. doi:https://doi.org/10.1093/genetics/163.2.663
2. Biswas S, El Kareh A, Qureshi M, Lee DMX, Sun CH, Lam JSH, Saw SM, Najjar RP (2024) The influence of the environment and lifestyle on myopia. J Physiol Anthropol 43(1):7. https://doi.org/10.1186/s40101-024-00354-7
3. Rodriguez NG, Claici AO, Ramos-Castaneda JA, González-Zamora J, Bilbao-Malavé V, de la Puente M, Fernandez-Robredo P, Garzón-Parra SJ, Garza-Leon M, Recalde S (2024) Conjunctival ultraviolet autofluorescence as a biomarker of outdoor exposure in myopia: a systematic review

and meta-analysis. Sci Rep 14(1):1097. https://doi.org/10.1038/s41598-024-51417-9
4. Methyl Cellulose, CAS number 9004-67-5. Sigma Aldrich. https://www.sigmaaldrich.com/deepweb/assets/sigmaaldrich/product/documents/756/144/m0387pis.pdf
5. Rodwell V, Patil M, Kuht HJ, Neuhauss SCF, Norton WHJ, Thomas MG (2024) Zebrafish optokinetic reflex: minimal reporting guidelines and recommendations. Biology 13(4). https://doi.org/10.3390/biology13010004
6. Scott CA, Marsden AN, Slusarski DC (2016) Automated, high-throughput, in vivo analysis of visual function using the zebrafish. Dev Dyn 245(5):605–613. https://doi.org/10.1002/dvdy.24398. Epub 2016 Mar 6

Chapter 14

Design and Validation of Mirtron-Mediated RNA Interference for Gene Therapy

Qingyun Zheng, Harry O. Orlans, Federica Staurenghi, Michelle E. McClements, and Robert E. MacLaren

Abstract

Mirtrons, a subclass of intronic microRNAs, offer an alternative RNA interference (RNAi) mechanism by bypassing Drosha processing. This protocol details the design and production of mirtron-based constructs aimed at achieving efficient gene knockdown. By embedding mirtrons in the 5′ untranslated region (UTR) of a transgene, we evaluated processing efficiency and RNAi activity. The compact size of mirtrons facilitates concurrent gene knockdown and supplementation within a single adeno-associated virus (AAV) vector. Our findings present a novel strategy for gene regulation in therapeutic and research applications.

Key words Mirtron, RNAi, DNA cloning, Gene therapy

1 Introduction

MicroRNAs (miRNAs) are pivotal posttranscriptional regulators that modulate gene expression by binding to complementary sequences on target messenger RNAs (mRNAs), leading to their degradation or translational repression [1]. Among these, mirtrons represent a unique class of short intronic miRNAs initially discovered in *Drosophila* and subsequently identified in mammals [2]. Unlike canonical miRNAs, which undergo processing by the microprocessor complex comprising Drosha and DGCR8, mirtrons circumvent Drosha cleavage. Instead, their maturation relies on splicing and lariat debranching mechanisms [2, 3]. This alternative biogenesis pathway not only underscores the versatility of RNA processing but also offers enhanced flexibility in designing synthetic applications for gene regulation.

Mirtrons function through the RNA interference (RNAi) pathway, similar to conventional miRNAs. After splicing, the mirtron forms a short hairpin structure, which is subsequently processed by Dicer to generate a mature miRNA duplex. One strand of the

Livia S. Carvalho (ed.), *Retinal Gene Therapy: Methods and Protocols*, Methods in Molecular Biology, vol. 3023, https://doi.org/10.1007/978-1-0716-5198-8_14,
© The Author(s), under exclusive license to Springer Science+Business Media, LLC, part of Springer Nature 2026

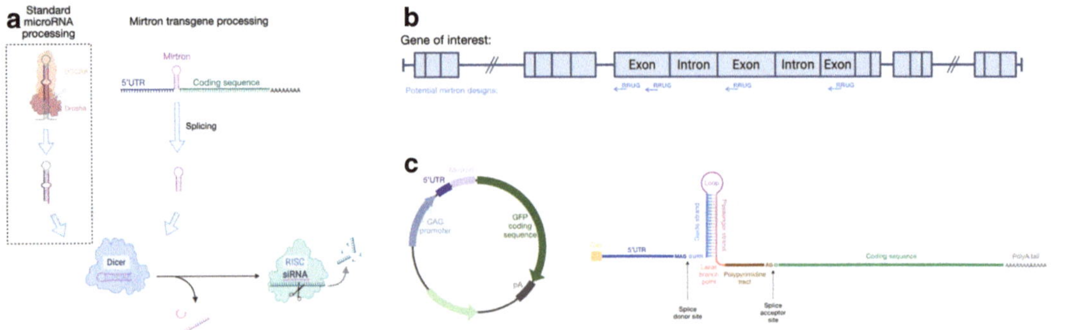

Fig. 1 (**a**) MicroRNA processing typically involves cleavage by Drosha and DGCR8 prior to Dicer processing to release the active small interfering RNA (siRNA) that binds to an mRNA target and induce translational repression or mRNA degradation via the RNA-induced silencing complex (RISC). Mirtrons bypass the need for Drosha/DGCR8. (**b**) Mirtrons require the GURR motif; therefore, designs are dictated by the presence of this motif on the target gene. (**c**) Mirtrons are inserted into a transgene within the 5′ untranslated region (UTR)

duplex is then loaded into the RNA-induced silencing complex (RISC), where it guides the complex to complementary target mRNAs, leading to gene silencing via translational repression or mRNA degradation (Fig. 1a) [3, 4]. Compared to canonical miRNAs, mirtrons offer several advantages, including a simplified processing mechanism and potential for efficient expression in diverse cellular environments [5].

Given their ability to mediate gene silencing, mirtrons have been explored as tools for gene knockdown in various research and therapeutic applications. A major challenge in gene therapy is achieving concurrent knockdown of a pathogenic transcript while simultaneously introducing a functional replacement gene [6]. Traditional approaches typically require two separate vectors: one for small RNA-mediated knockdown and another for transgene expression. This dual-vector requirement complicates experimental setups and makes the balance of knockdown and replacement difficult to achieve through vector dosage.

To address this limitation, we present a novel strategy for designing, generating, and validating mirtron-based vectors for gene regulation. By integrating single- or double-mirtron constructs into the 5′ untranslated region (UTR) of a transgene within a single vector, we enable simultaneous target gene silencing and transgene expression without compromising downstream gene function. The compact size of mirtrons allows for efficient packaging within an adeno-associated virus (AAV) vector, streamlining gene therapy applications. This approach ensures stable expression while achieving effective gene knockdown, making mirtrons a promising tool for therapeutic gene modulation and research applications.

2 Materials

2.1 Molecular Cloning

1. Restriction enzyme: EcoRV-HF.
2. Any commercially available 10× oligo annealing buffer.
3. Any commercially available PCR Purification Kit such as the QIAquick PCR Purification Kit.
4. Nuclease-free water.
5. Any commercially available DNA ligation kit: T4 DNA ligase, T4 ligase buffer.
6. Competent *E. coli* cells.
7. SOC medium: 20 g/L tryptone, 5 g/L yeast extract, 0.5 g/L sodium chloride, 0.186 g/L potassium chloride, 1 M glucose, 2 M magnesium chloride. Prepare 0.8 L of H2O in a suitable container. Add 20 g of tryptone to the solution. Add 5 g of yeast extract to the solution. Add 0.5 g of sodium chloride to the solution. Add 0.186 g of potassium chloride to the solution. Adjust pH with NaOH solution to 7.0. Add H2O until the volume is 1 L. After autoclaving, add 20 mL filtered 1 M glucose and 5 mL autoclaved 2 M magnesium chloride.
8. Luria-Bertani (LB) agar: 25 g LB Broth (Miller), 15 g agar, 100 μg/mL ampicillin. Add 25 g of LB Broth and 15 g of agar to 1 L of distilled water. Mix well by inverting bottle several times until powder is dissolved and sterilize by autoclaving for 20 min at 15 psi (1.05 kg/cm^2) on liquid cycle. When solution is still liquid cool enough to safely handle bottle, supplement with 100 μg/mL ampicillin. Pour agar into 60 mm plates within a laminar flow hood. Allow ~50 mL per plate. Leave plates in laminar hood until agar sets (~30 mins). When the agar has set, replace lid, invert plates, and store in an airtight bag at 4 °C.
9. LB base solution: 10 g tryptone, 10 g sodium chloride (NaCl), 5 g yeast extract. In 950 mL of Milli-Q H2O, add 10 g tryptone, 10 g sodium chloride (NaCl), and 5 g yeast extract. Mix until powder is dissolved. pH solution to ~7.0 using sodium hydroxide (NaOH). Adjust solution to a final volume of 1000 mL. Autoclave using liquid cycle.
10. Reverse osmosis (RO) water.
11. Any commercially available Plasmid Miniprep Kit.

2.2 In Vitro Validation Assay

1. Human Embryonic Kidney 293 (HEK293) cells.
2. Culture medium: Complete Dulbecco's Modified Eagle Medium (DMEM), 10% filtered fetal bovine serum, 1% penicillin–streptomycin.

3. 3.Transfection reagents such as or equivalent to Lipofectamine 3000.
4. Any commercially available Luciferase Assay Kit such as or equivalent to Dual-Glo® Luciferase Assay System (Promega).
5. Multilabel plate reader.
6. Black plastic 96-well plates.
7. 12-well plates.
8. 96-well plate.
9. Fluorescence microscope.

2.3 Splicing Efficiency Assessment

1. Any commercially available RNA extraction kit.
2. Any commercially available reverse-transcription kit.
3. Any commercially available PCR Master Mix such as or equivalent to the Platinum™ SuperFi™ II PCR Master Mix (Thermo Fisher Scientific).
4. Any commercially available DNA gel loading dye.
5. 1% agarose gel: Agarose, 1× TBE buffer. Weigh out 1 g of agarose powder and add it to 100 mL of 1× TBE in a microwaveable flask. Heat the mixture in a microwave for 1–3 min, swirling the flask between intervals, until the agarose is completely dissolved. Allow the solution to cool to a temperature where you can comfortably hold the flask (around 50 °C). Add the fluorescent nucleic acid stain at the recommended amount. Pour the cooled, dissolved agarose solution into a gel tray with a comb in place to create wells.
6. Any commercially available fluorescent nucleic acid stain.
7. 100 bp DNA ladder.

3 Methods

3.1 Mirtron Design

Mirtrons begin with the sequence GURR (with R being either G or A); this motif can therefore be used to screen the target gene for potential mirtron binding regions (Fig. 1b). The guide region continues from the GURR motif, followed by a loop element and the passenger sequence, which is complementary to the guide. A lariat branch point and a polypyrimidine tract complete the mirtron structure (Fig. 1c). Our previous research [8] identified that inclusion of exonic splice enhancer sites (ESEs) on either side of the mirtron design improves splicing efficiency; therefore, these are also recommended (*see* **Note 1**).

Example of Mirtron Construct Design

Synthesize mirtron sequences with optimized flanking regions (ESEs) to ensure efficient splicing. Design single-mirtron (e.g., Mirtron3, Mirtron5) and double-mirtron combinations (M3/3, M3/5) with appropriate flanking sequences.

Optimized flanking regions are used in our constructs: upstream 5′ flank TGACGGGAACTACAAGACCCGCGCT GAAGTCAAGTTCGAAG and downstream 3′ flank GTGA CACCCTGGTGAATAGAATCGAGCTGAAGGGCATTGACTT TAAGGAGGAT [7, 8] (*see* **Note 1**).

We previously designed seven mirtron sequences for screening, of which the selected top two were as follows [8] (*see* **Note 2**):

Mirtron3(5′-3′): GTGAGGAAGTTGATGGGGAAGTTCAAGA GAACTTCCCCATCAACTTCCTCA CAGTTTTTTCTCTTTCTTTT;

Mirtron5(5′-3′): GTAGATGACAAAAGACTCGTTTTCAAGA GAAACGAGTCTTTTGTCATTTCT CAGTTTTTTCTCTTTCTTTT.

3.2 Cloning of Single and Double Mirtrons into the 5'UTR of the Reporter Gene

For screening mirtron designs, we used a CAG.GFP reporter plasmid and inserted the mirtrons in a 5'UTR sequence between the CAG and GFP elements via an EcoRV site.

1. Vector preparation: Digest the CAG-GFP vector with EcoRV-HF to obtain a linearized backbone with blunt ends. The digestion reaction (20 μL): 1 μg of CAG-GFP plasmid, 1 μL EcoRV-HF enzyme, 2 μL 10× CutSmart buffer, and nuclease-free water up to 20 μL. Incubation was performed at 37 °C (*see* **Note 3**).
2. Vector purification: Purify the digested vector using the QIAquick PCR Purification Kit to eliminate residual enzymes and undigested plasmids.
3. Elute the DNA in 30 μL nuclease-free water and quantify using a NanoDrop spectrophotometer.
4. Synthesis of the mirtron construct: Order synthesized single and double mirtrons plus flanking sequences as two complementary single-stranded oligonucleotides, each containing EcoRV-compatible blunt ends to enable direct ligation into the linearized vector (*see* Table 1). DNA fragment synthesis can be sourced from many companies, for example, IDT and Twist Biosciences.
5. Annealing of synthetic mirtron oligonucleotides: To generate a double-stranded mirtron insert, anneal the complementary oligonucleotides: Annealing reaction (20 μL total volume): 1 μL (100 μM) forward oligo, 1 μL (100 μM) reverse oligo, 2 μL 10× annealing buffer (100 mM NaCl, 10 mM Tris–HCl pH 7.5, 1 mM EDTA), and 16 μL nuclease-free water. Mix

Table 1

Example sequences of mirtron oligos for annealing [8]. ESE flanking regions are in italics; mirtron sequences are bold

Oligo name	Oligo sequence (5′-3′)
M3-FW	GATATC*GAATTCTAGCTTGATTGACGGGAACTACAAGACCCGCGCTGAAGTCAAGTT*CGAAG**GTGAGGAAGTTGATGGGGAAGTTCAAGAGAACTTCCCCATCAACTTCCTCACAGTTTTTCTCTTTCTTTT**CGAAGGT*GACACCCTGGTGAATAGAATCGAGCTGAAGGGCATTGACTTTAAGGAG*GATATC
M3-RC	GATATC*CTCCTTAAAGTCAATGCCCTTCAGCTCGATTCTATTCACCAGGGTGTCAC*CTTCG**AAAAGAAAGAGAAAAAACTGTGAGGAAGTTGATGGGGAAGTTCTCTTGAACTTCCCCATCAACTTCCTCAC**CTTCG*AACTTGACTTCAGCGCGGGTCTTGTAGTTCCCGTCAATCAAGCTAGAATTC*GATATC
M5-FW	GATATC*GAATTCTAGCTTGATTGACGGGAACTACAAGACCCGCGCTGAAGTCAAGTT*CGAAG**GTAGATGACAAAAGACTCGTTTTCAAGAGAAACGAGTCTTTTGTCATTTCTCAGTTTTTCTCTTTCTTTT**CGAAG*GTGACACCCTGGTGAATAGAATCGAGCTGAAGGGCATTGACTTTAAGGAG*GATATC
M5-RC	GATATC*CTCCTTAAAGTCAATGCCCTTCAGCTCGATTCTATTCACCAGGGTGTCAC*CTTCG**AAAAGAAAGAGAAAAAACTGAGAAATGACAAAAGACTCGTTTCTCTTGAAAACGAGTCTTTTGTCATCTAC**CTTCG*AACTTGACTTCAGCGCGGGTCTTGTAGTTCCCGTCAATCAAGCTAGAATTC*GATATC
M3/3-FW	GATATC*GAATTCTAGCTTGATTGACGGGAACTACAAGACCCGCGCTGAAGTCAAGTT*CGAAG**GTGAGGAAGTTGATGGGGAAGTTCAAGAGAACTTCCCCATCAACTTCCTCACAGTTTTTCTCTTTCTTTT**CGAAG*GTGACACCCTGGTGAATAGAATCGAGCTGAAGGGCATTGACTTTAAGGAGGATCTGCAGTGACGGGAACTACAAGACCCGCGCTGAAGTCAAGTT*CGAAG**GTGAGGAAGTTGATGGGGAAGTTCAAGAGAACTTCCCCATCAACTTCCTCACAGTTTTTCTCTTTCTTTT**CGAAG*GTGACACCCTGGTGAATAGAATCGAGCT*GATATC
M3/3-RC	GATATC*AGCTCGATTCTATTCACCAGGGTGTCACC*TTCG**AAAAGAAAGAGAAAAAACTGTGAGGAAGTTGATGGGGAAGTTCTCTTGAACTTCCCCATCAACTTCCTCA**CCTTCG*AACTTGACTTCAGCGCGGGTCTTGTAGTTCCCGTCACTGCAGATCCTCCTTAAAGTCAATGCCCTTCAGCTCGATTCTATTCACCAGGGTGTCA*CCTTCG**AAAAGAAAGAGAAAAAACTGTGAGGAAGTTGATGGGGAAGTTCTCTTGAACTTCCCCATCAACTTCCTCAC**CTTCG*AACTTGACTTCAGCGCGGGTCTTGTAGTTCCCGTCAATCAAGCTAGAATTC*GATATC
M3/5-FW	GATATC*GAATTCTAGCTTGATTGACGGGAACTACAAGACCCGCGCTGAAGTCAAGTT*CGAAG**GTGAGGAAGTTGATGGGGAAGTTCAAGAGAACTTCCCCATCAACTTCCTCACAGTTTTTCTCTTTCTTTT**CGAAG*GTGACACCCTGGTGAATAGAATCGAGCTGAAGGGCATTGACTTTAAGGAGGATCTGCAGTGACGGGAACTACAAGACCCGCGCTGAAGTCAAGTT*CGAAG**GTAGATGACAAAAGACTCGTTTTCAAGAGAAACGAGTCTTTTGTCATTTCTCAGTTTTTCTCTTTCTTTT**CGAAG*GTGACACCCTGGTGAATAGAATCGAGCT*GATATC

(continued)

Table 1
(continued)

Oligo name	Oligo sequence (5′-3′)
M3/5-RC	GATATC*AGCTCGATTCTATTCACCAGGGTGTCACC* TTCG**AAAAGAAAGAGAAAAAACTGAGAAATGACAAAAGACTCGTTTCTC** **TTGAAAACGAGTCTTTTGTCATCTAC**CTTCG*AACTTGACTTCAGCGCGGG* *TCTTGTAGTTCCCGTCACTGCAGATCCTCCTTAAAGTCAATGCCCTTCAGC* *TCGATTCTATTCACCAGGGTGTCACC*TTCG**AAAAGAAAGAGAAAAAACTG** **TGAGGAAGTTGATGGGGAAGTTCTCTTGAACTTCCCCATCAACTTCC** **TCAC**CTTCG*AACTTGACTTCAGCGCGGGTCTTGTAGTTCCCGTCAATCAAGC* *TAGAATTC*GATATC

gently and then heat to 95 °C for 5 min in a thermocycler. Gradually reduce to room temperature overnight to allow base pairing (*see* **Note 4**).

6. Ligation of annealed mirtron into the vector: Prepare a standard ligation reaction (20 μL total volume): 50 ng EcoRV-digested CAG-GFP vector, 3:1 molar ratio of annealed mirtron insert, 1 μL T4 DNA ligase, 2 μL 10×T4 ligase buffer, and nuclease-free water up to 20 μL. Various incubation protocols are acceptable (*see* **Note 5**).
7. Heat-inactivate the ligation mixture at 65 °C for 10 min before transformation.
8. Transformation into competent *E. coli*: Add 5 μL of the ligation mix to 50 μL chemically competent DH5α *E. coli* cells. Incubate the tube on ice for 30 min, heat-shock at 42 °C for 45 s (*see* **Note 6**), and then place on ice for 5 min (*see* **Note 7**). Add 500 μL SOC medium and incubate the cell mixture at 37 °C for 1 h with shaking at 250 rpm.
9. Plating and colony selection: Spread 100 μL transformed cells on LB-agar plates supplemented with 100 μg/mL ampicillin. Incubate the plates at 37 °C overnight to allow colony formation.
10. Pick individual colonies for further growth. Pierce each with a sterile pipette tip and use this to inoculate 3 mL LB with 100 μg/mL ampicillin at 200 rpm for 12 h.
11. Use the Qiagen Miniprep Kit to extract the plasmid DNA.
12. Sequence validation: To confirm the correct insertion and orientation of the mirtron, 10 μL of plasmid DNA was subjected to Sanger sequencing using the reverse primer 5′-GTGTTTAAAAGACAGTCG.

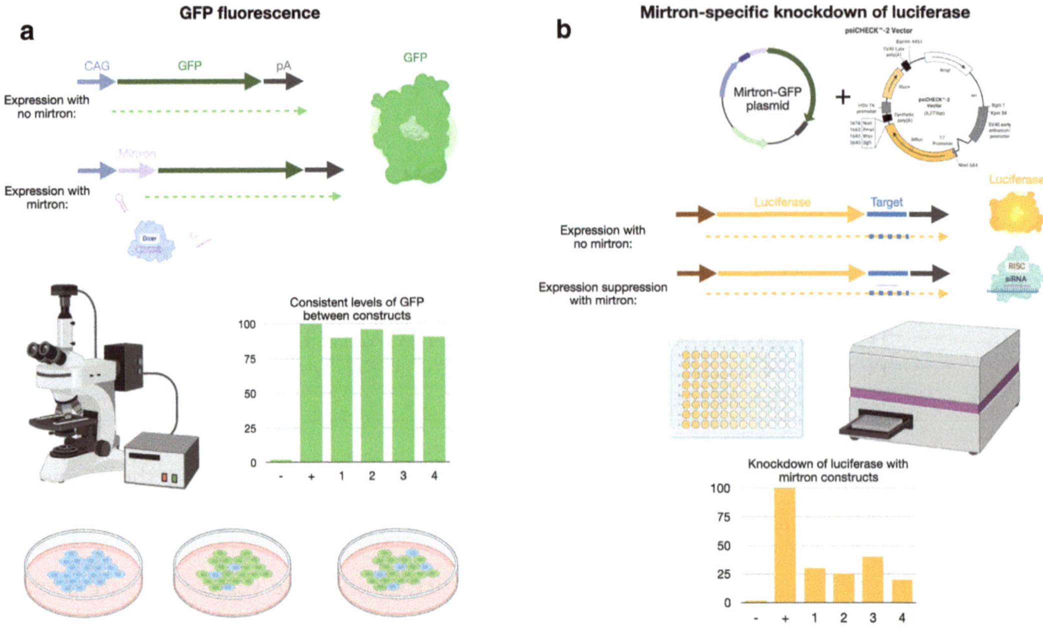

Fig. 2 In vitro assessments of mirtron constructs. (**a**) The impact of mirtrons within the 5'UTR on downstream transgene expression can be measured by transfecting HEK293 cells and imaging GFP fluorescence 48 h post-transfection. The presence of mirtrons should not significantly impact GFP levels compared to no mirtron controls. (**b**) A Dual-Glo® assay can be performed to compare the efficiency of knockdown for each mirtron. By co-transfecting with a PsiCHECK2 construct containing the target sequence, *Renilla* luciferase levels will be reduced if mirtron splicing and targeting is successful

3.3 Functional Verification Using In Vitro Assays

In vitro assays can be performed to (1) confirm mirtron presence in the 5'UTR does not adversely impact expression of the downstream transgene, (2) confirm successful splicing of the mirtrons, and (3) assess the ability of each mirtron to knock down expression of its target. To achieve the final objective, the target sequence can be cloned into the PsiCHECK2 plasmid (C8021, Promega). By inserting the target of interest downstream of the *Renilla* luciferase, miRNA knockdown of the target sequence will knock down *Renilla*. For all in vitro assays, perform each sample in triplicate and repeat each assay at least three times.

1. To assess for the impact of including mirtrons in the 5'UTR on downstream transgene expression, HEK293 cells can be seeded into 96-well culture plates. After 24 h, transfect with 65 ng mirtron-GFP plasmids, using CAG-GFP as the control.
2. After 48 h, image the cells to determine GFP fluorescence (Fig. 2a).
3. To assess for target knockdown, co-transfect HEK293 cells with 65 ng of each mirtron-GFP plasmid and 32.5 ng PsiCHECK2 plasmid containing the target sequence to perform a Dual-Glo® assay.

4. After 48 h, remove 30 μL of media from each well.
5. Add 70 μL of *Renilla* luciferase substrate to each well and incubate the plate for 20 min.
6. Record the levels of *Renilla* luminescence signal, for example, using a FLUOstar Omega plate reader (BMG Labtech).
7. Add 70 μL of Stop & Glo reagent to each well and incubate the plate at room temperature for 20 min.
8. Record the levels of firefly luminescence. The firefly luminescence is an internal control within the PsiCHECK2 plasmid and will not be targeted by the mirtrons.
9. Determine mirtron-mediated mRNA knockdown by first adjusting for background luminescence for both sets of data (using readings from untransfected wells) and then calculate the relative luciferase ratio *Renilla*/firefly for each sample. Normalize to readings from samples transfected with the mirtronless CAG-GFP plasmid (Fig. 2b).

3.4 Splicing Efficiency of Artificial Mirtrons

1. Seed HEK293 cells into 12-well plates and transfect 24 h later with 1 μg CAG-GFP or mirtron-GFP constructs.
2. After 48 h, image the cells to confirm cell health and then harvest for RNA extraction (*see* **Note 8**).
3. Pellet cells by centrifugation (500 ×g) for 1 min. Discard supernatant.
4. Resuspend each pellet in 300 μL of RNA lysis buffer by pipetting gently to avoid foaming.
5. Transfer up to 800 μL to a gDNA removal column fitted with a collection tube and label collection tubes.
6. 16,000×g, spin for 30 s to remove most of the gDNA. Keep the flow-through (RNA partitions here). Discard the gDNA removal column.
7. Add an equal volume of ethanol (≥ 95%) to the flow-through and mix thoroughly by pipetting. Do not vortex. To exclude RNA ≤ 200 nt, add only 1/2 volume ethanol to flow-through. The addition of ethanol creates favorable conditions for RNA to bind to the RNA purification column.
8. Transfer the mixture to an RNA purification column fitted with a collection tube. Spin for 30 s. Discard flow-through.
9. dd 500 μL RNA wash buffer and spin for 30 s. Discard flow-through. This ensures all salts are removed prior to the addition of DNase I.
10. To remove residual gDNA, combine 5 μL DNase I with 75 μL DNase I Reaction Buffer per sample and apply the mixture directly to the top of the matrix. Incubate for 15 min at room temperature.

11. Add 500 μL RNA priming buffer and spin for 30 s. Discard flow-through.
12. Add 500 μL RNA wash buffer and spin for 30 s. Discard flow-through.
13. Add another 500 μL RNA wash buffer and spin for 2 min. Transfer the column to an RNase-free microfuge tube.
14. Add 100 μL nuclease-free water directly to the center of column matrix and spin for 30 s. Place RNA on ice if being used immediately for the subsequent steps or at −80 °C for long-term storage.
15. Determine the concentration and quality of your RNA, preferably using a Qubit, but a NanoDrop is acceptable (both available from Thermo Fisher Scientific).
16. Convert RNA to complementary DNA (cDNA) by preparing the reverse-transcription reaction solution on ice.
17. In PCR tubes, prepare for each sample: 2 μL 5×PrimeScript RT Master Mix, 500 ng total RNA, and RNase Free dH2O to top up the final volume to 10 μL.
18. Gently mix each reaction solution.
19. Place the samples in a PCR machine and run: 37 °C 15 min (reverse-transcription), 85 °C 5 sec (heat inactivation of reverse transcriptase), and 4 °C (hold until ready to collect samples).
20. Perform PCR using primers spanning the transcription start site (TSS) to the GFP coding sequence. Prepare the PCR solution on ice: 2× Platinum™ SuperFi™ II PCR Master Mix 10 μL, mirtron-spanning primers (forward primer 0.5 μM, 1 μL: 5'-GCTGGTTATTGTGCTGTCTC; reverse primer 0.5 μM, 1 μL: 5'-GGGCTGCAGGAATTCGAT). cDNA: 1 μL, nuclease-free water 7 μL.
21. Gently mix each reaction solution.
22. Place in the PCR machine and run a three-step protocol, for example, initial denaturation at 98 °C for 30 s, followed by 30 cycles of denaturation at 98 °C for 5 s, annealing at 60 °C for 10 s, and extension at 72 °C for 10 s, and finish with a final extension phase at 72 °C for 5 min. Place samples in the fridge if not moving immediately to the next step.
23. Assess the splicing products by gel electrophoresis: Add 4 μL 6× loading buffer to the full volume of each PCR reaction and then load on a 1% agarose gel and run at 100 V for at least 1 h (Fig. 3).
24. Carefully cut out individual bands for each sample and place each gel piece in its own 1.5 mL tube.
25. Extract the amplicon from each gel piece, for example, using the QIAquick Gel Extraction Kit and following the manufacturer's standard protocol (28,704, QIAGEN).

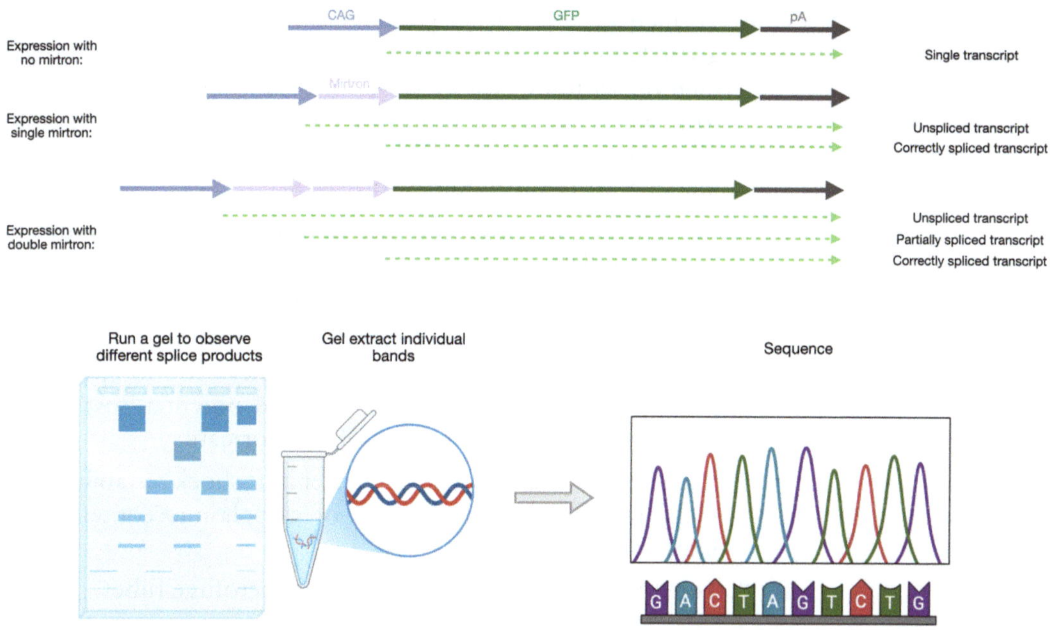

Fig. 3 Assessment of mirtron splicing efficiency. Using a forward primer binding downstream of the transcriptional start site and a reverse primer that binds in the transgene coding sequence, mRNA transcripts can be amplified across the 5′UTR. When no mirtron is present, a single transcript and therefore a single band is expected. Incomplete or partial splicing of constructs containing mirtrons will lead to multiple bands, which can be gel extracted and sent for sequencing

26. Assess the quality and concentration of each purified sample and then send for Sanger sequencing with your preferred service provider.

Combined, the steps outlined in this protocol should enable the design, creation, and assessment of mirtrons for targeted gene knockdown.

4 Notes

1. These flanking sequences need to be identified using the RESCUE-ESE tool, which predicts exonic splicing enhancer (ESE) motifs. Also, these flank lengths were determined using in silico splicing analysis (SplicePort) to maximize donor site strength while avoiding unintended "ATG" sequences that could introduce alternative translation initiation.
2. Candidate mirtron sequences need to be analyzed using SplicePort and Human Splice Finder to identify those with strong donor motifs and then using the Whitehead siRNA prediction tool with RNAiRNA interference (RNAi) efficiency evaluation.
3. To optimize blunt-end vector digestion efficiency, incubate the reaction for 4 h at the optimal temperature to improve cleavage efficiency while minimizing potential star activity.

4. To increase the efficiency of annealing of complementary oligonucleotides, mix equimolar amounts of sense and antisense oligonucleotides, heat to 95 °C for 5 min, and then slowly cool to room temperature to ensure proper annealing. A thermocycler with a gradual cooling ramp (1 °C per min) is recommended for optimal duplex formation.
5. Ligase reaction should be carried out at 16 °C overnight to enhance efficiency.
6. The optimal heat shock time is 42 °C for 45 s; too short (<30 s) would lead to insufficient membrane permeability, leading to lower transformation efficiency. Too long (>50 s) would increase membrane damage and potential loss of cell viability, reducing the number of transformants.
7. Placing cells on ice immediately after heat shock restabilizes the membrane, preventing leakage of cytoplasmic contents and improving cell survival.
8. For RNA extraction, use RNase-free microfuge tubes.

Acknowledgments

This work was supported by grants from the Medical Research Council UK, NIHR Oxford Biomedical Research Centre UK and Medical and Life Sciences Translational Fund UK.

Conflict of Interest The authors have no conflicts of interest to declare that are relevant to the content of this chapter.

References

1. Ying SY, Chang DC, Lin SL (2008) The microRNA (miRNA): overview of the RNA genes that modulate gene function. Mol Biotechnol 38(3): 257–268
2. Okamura K, Hagen JW, Duan H et al (2007) The mirtron pathway generates microRNA-class regulatory RNAs in Drosophila. Cell 130(1): 89–100
3. Westholm JO, Lai EC (2011) Mirtrons: microRNA biogenesis via splicing. Biochimie 93(11): 1897–1904
4. Flynt AS, Greimann JC, Chung WJ et al (2010) MicroRNA biogenesis via splicing and exosome-mediated trimming in Drosophila. Mol Cell 38(6):900–907
5. O'Brien J, Hayder H, Zayed Y et al (2018) Overview of microRNA biogenesis, mechanisms of actions, and circulation. Front Endocrinol (Lausanne) 9:402
6. Xu Y, Li Z (2020) CRISPR-Cas systems: overview, innovations and applications in human disease research and gene therapy. Comput Struct Biotechnol J 18:2401–2415
7. Fairbrother WG, Yeh RF, Sharp PA et al (2002) Predictive identification of exonic splicing enhancers in human genes. Science 297(5583): 1007–1013
8. Orlans HO, McClements ME, Barnard AR et al (2021) Mirtron-mediated RNA knockdown/replacement therapy for the treatment of dominant retinitis pigmentosa. Nat Commun 12(1): 4934

Part IV

Clinical Validation of Retinal Gene Therapies

Chapter 15

Applications of Fundus-Tracked Perimetry in the Assessment of Visual Sensitivity

Alexis Ceecee Britten-Jones, Janise Dagelet, and Lauren N. Ayton

Abstract

Fundus-tracked perimetry, also called microperimetry, is increasingly used as an outcome measure in retinal gene therapy clinical trials. By combining static automated perimetry and real-time fundus tracking, fundus-tracked perimetry enables accurate assessment of visual sensitivity in the central retina to precisely identify areas of impaired or preserved function.

This chapter describes the clinical methods used to perform fundus-tracked perimetry and strategies to improve the reliability of fundus-tracked perimetry measurements, including operator training, patient preparation, and test monitoring.

Key words Fundus-tracked perimetry, Microperimetry, Retinal sensitivity, Visual function, Macular sensitivity, Perimetry

1 Introduction

Perimetry is a clinical technique for measuring a person's visual function, by assessing retinal sensitivity at different parts of the visual field. Perimetry methods include static and kinetic perimetry. Static perimetry measures a person's visual sensitivity around a point of fixation, by detection of light stimuli presented at different points around the visual field. In kinetic perimetry, a light stimulus is moved from a non-seeing area to a seeing area, and the location where the light stimulus is first seen is recorded.

Fundus-tracked perimetry (also termed microperimetry) combines static automated perimetry with real-time fundus tracking. This technique measures a person's response to light stimuli at various predetermined retinal locations around a point of fixation and superimposes this data onto a retinal image captured by fundus photography or scanning laser ophthalmoscopy. Using the retinal image to track the area of testing, fundus-tracked perimetry enables spatially resolved mapping of visual sensitivity in the central retina

Livia S. Carvalho (ed.), *Retinal Gene Therapy: Methods and Protocols*, Methods in Molecular Biology, vol. 3023,
https://doi.org/10.1007/978-1-0716-5198-8_15,
© The Author(s), under exclusive license to Springer Science+Business Media, LLC, part of Springer Nature 2026

to precisely identify areas of impaired or preserved function. This can be done even in patients with unstable fixation or extrafoveal fixation, such as age-related macular degeneration and Stargardt disease.

Fundus-tracked perimetry has been used in many observational research studies to assess structure–function correlations and to monitor the progression of retinal diseases, including age-related macular degeneration [1, 2], diabetic retinopathy [3, 4], glaucoma [5, 6], and inherited retinal diseases [7–9]. Fundus-tracked perimetry has also been used as a secondary outcome measure in several retinal gene therapy clinical trials [10], including clinical trials evaluating subretinal gene therapy treatment for choroideremia [11–13], X-linked retinitis pigmentosa [14], and achromatopsia [15]. Outcomes derived from fundus-tracked perimetry testing that have been used in retinal gene therapy clinical trials include change in mean sensitivity and change in maximal point sensitivity [11, 13, 14, 16, 17], fixation stability [15], and number of scotomas [18, 19].

Fundus-tracked perimetry also has broader applications in clinical practice. For example, it can be used by low-vision specialists in coaching their patients with significant vision loss in making the most effective use of their remaining vision, by identifying their preferred retinal locus (PRL).

There are several commercially available fundus-tracked perimetry devices. The Microperimeter-1 (MP-1, NIDEK Technologies Srl, Padova, Italy), released in 2003, was the first fundus-tracked perimeter that was made commercially available. The MP-1 was subsequently superseded by the MP-3 (2015, NIDEK CO., Ltd., Gamagori, Japan), which has a 45° testing field and a range of stimulus intensity between 0 and 36 decibels (dB). Goldmann I to V stimuli sizes are available on the MP-3. The MP-3 is also presently the highest eye tracker frequency available on commercial devices (30 Hz).

The Macular Integrity Assessment (MAIA, CenterVue SpA., Icare Finland) was released in 2009. The MAIA has a 36° testing field; the range of stimulus intensity is between 0 dB and 20 dB, and the eye tracker frequency is 25 Hz. Inter-device comparison from healthy eyes showed that retinal sensitivity measures higher and luminance and contrast sensitivity measure lower for MAIA-generated values compared with the MP-3; however, the values were consistently correlated, and standardized conversion factor could be applied to compare results from the two instruments [20]. The MAIA uses a Goldmann III light stimuli (diameter of 0.43°).

Endpoints in gene therapy clinical trials need to be valid and reproducible to accurately assess the treatment efficacy. The reliability of fundus-tracked perimetry measurements is affected by several factors, including operator training and instructions, patient preparation, testing procedure, and monitoring.

In a comprehensive review by Pfau et al., the authors recommend several key procedural points that can improve the accuracy of fundus-tracked perimetry testing [21]:

1. Adequate training of the operator and instruction and practice for the patient.
2. Proper test setup and image focusing.
3. Ten-minute adaptation to ambient lighting before testing (particularly important following intense light exposure, such as retinal imaging).
4. Consistency in dilation state between visits in a clinical trial setting.
5. Ensuring correct grid placement at the macula or PRL.
6. Careful monitoring of the reliability factors of the test and continuous instruction to correct any errors that the patient may make during testing.

In this chapter, we will describe the procedures to perform fundus-tracked perimetry examination using the MAIA device in undertaking fundus-tracked perimetry examination. The MAIA device has been used in several more recent gene therapy clinical trials in inherited retinal diseases [10, 14, 22].

2 Materials

1. The Macular Integrity Assessment (MAIA, CenterVue SpA., Icare Finland) unit.
2. Alcohol for disinfecting the chin rest and forehead strap.
3. An eye patch to test each eye individually.
4. Pupil dilation ophthalmic solution, if used (e.g., 0.5% or 1% tropicamide or 2.5% phenylephrine ophthalmic solution. *See* **Note 1**).
5. Custom grid pattern (.xml file), if a custom test grid is used (*see* **Notes 2** and **3**).

3 Methods

3.1 Test Preparation

1. Testing should be performed in a quiet room, free of distractions.
2. The standard MAIA is set for a background illumination of 1.27 cd/m^2 for mesopic testing (*see* **Notes 4** and **5**).

3.2 Instrument Setup and Calibration

1. Switch on the instrument, and, after the software loads, the machine may prompt a calibration check to ensure that the fixation target is within the correct viewing area (Fig. 1).
2. Calibration can be confirmed by looking through the front lens (sitting in the patient position) to check that the position of the white spot is within the red fixation target.
3. If it is, select "OK" on the operator screen. If the target is not within the red circle, use the focus buttons to calibrate the machine.

3.3 Selecting a Patient

3.3.1 Add a New Patient

1. Select the "New Patient" button and enter the last name, first name, patient ID, and date of birth (mandatory fields).
2. Select completed and the test screen will open with the new patient record.

3.3.2 Selecting an Existing Patient

1. Select patient from the start-up screen or enter patient details into the search bar by last name, first name, or patient ID.
2. The arrow in the header allows the operator to sort by ID, last name, date of birth, or date of last visit.

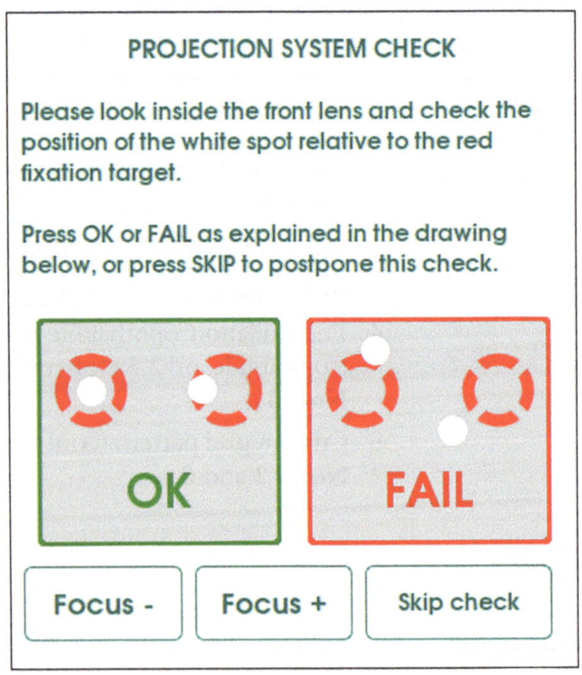

Fig. 1 Instrument calibration for the Macular Integrity Assessment (CenterVue SpA., Icare Finland) fundus-tracked perimeter

3.4 Patient Instructions

3.4.1 Purpose of the Test

1. The purpose of the test should be explained to the patient. For example, this test is to assess their central and peripheral vision by their ability to detect lights at different parts of their vision while they fixate on a steady target or object.

3.4.2 Fixation Target and Test Stimuli

1. Instruct the patient that they will see a red circle in the center of their vision, which they should fixate on throughout the entire test. Small flashes of white light will appear at different parts of their vision. The patient is then asked to click the button on the buzzer every time they detect a flash of white light while continuing to fixate on the red circle.
2. The patient should be reassured that the instinct to look toward the light flash is natural but to do their best to concentrate on the red circle and to not look at the white flashing lights in other parts of their vision as they appear.
3. Inform the patient that the location of the white light stimuli could be further out to the periphery or closer to the central red circle. The lights could vary in intensity of brightness, and a light in the same location could be tested several times.
4. There may be occasions where no stimuli are seen for 10–15 s (possibly longer in patients with large scotomas or areas of declined retinal sensitivity). Patients should be reminded that this is okay and to continue fixating on the red circle even if stimuli is not being seen consistently.
5. If a stimulus is missed accidentally, the patient should not be overly concerned as each point may be retested several times.

3.4.3 Blinking

1. Remind the patient that they should blink as normal. Blink rates tend to decrease during tasks that require concentration, which can affect both and tear film stability and test results (*see* **Note 6**).

3.4.4 Rest and Pause

1. According to the manufacturers' instructions, a standard MAIA examination using the 4–2 strategy has an average duration of 5.5 min (*see* **Note 7**). However, depending on the grid used, retinal pathology, fixation, and patient fatigue, each test can last more than 10 min in duration, resulting in a combined duration of 20 min or more for both eyes.
2. Since each individual test can exert strain on the patient's concentration and fatigue can affect the reliability of the testing,[12] the patient should be given the opportunity to take breaks mid-assessment.
3. If a break is desired, the patient should verbally inform the operator that they want to take a break and the operator will pause the test, which will stop the test timer and stimuli. The patient can also hold down the button on the response unit to pause the test.

3.5 Test Selection and Demonstration

1. The standard test in MAIA fundus-tracked perimetry is the "Expert Test."
2. Alternatively, "Fast Test" is a suprathreshold strategy that measures two levels of sensitivity (27 dB and 25 dB), designed to screen patients with no known retinal disease.
3. A demonstration test should be performed first to familiarize the patient with the testing procedure. This can be done using the desired testing grid.

3.6 Patient Setup

1. Patch the non-testing eye and seat the patient comfortably with their chin on the chin rest. The patient should not be slouching or straining their neck.
2. The patient's position can be adjusted using the seat height, machine height, and the chin rest dial, to make sure that their eye lines up with the eye symbol on the headrest.
3. Give the patient the response unit and ask the patient to demonstrate pressing the push button.

3.7 Focusing and Aligning the Fundus Image

1. Select "Go to OD" or "Go to OS," as determined by the eye being tested (i.e., non-patched eye).
2. Ask the patient to look for a small red circle straight ahead, which is the fixation target. The red fixation circle has a white light in the middle.
3. If the patient is unable to see the small red circle, the larger red circle can also be turned on using the "Toggle Big Circle" button. This can help participants with poor central vision to locate and fixate on a target by using their eccentric fixation.
4. Select "auto-alignment" or manually align the instrument until a light spot showing the retina can be seen on the testing screen.
5. Zoom into the retina using the "Forward" (toward patient) or "Backward" (away from patient) controls until the fundus image populates the entire view of the testing screen (or as much of the screen as possible).
6. The position of the camera can be further adjusted using the joystick/toggle functions of the MAIA perimeter. The joystick/toggle moves the machine rather than the patient; therefore, toggling left on the screen will move the camera to the examiner's left and the patient's right.
7. Adjust the focus of the fundus image manually using the "Focus +"/"Focus −" buttons, or the focus will automatically adjust after pressing the "Start" button. Brightness will also be adjusted automatically during this process.

8. The optimal alignment and focus should position the point of fixation approximately in the middle of the screen with the optic nerve head bordering the edge of the image screen. The retinal veins and fovea should be as sharply defined as possible.

3.8 Grid Selection

1. Select the desired test grid using the "Change Grid" option (*see* **Notes 3** and **4**). The test will generally default to the last test performed.
2. Select testing strategy (*see* **Note 7**). Available strategies include:
   (a) *Strategy 4–2*: Perimetric standard where the light intensity changes in 4 dB steps until there is a change from not seen to seen (or from seen to not seen) and then in 2 dB steps until the response is reversed. The full-threshold 4–2 strategy is the standard strategy used in clinical studies and clinical trials.
   (b) *Four-level fixed*: Suprathreshold strategy where four different stimuli in intensities are projected: 25 dB, 15 dB, 5 dB, and 0 dB. This strategy provides an initial assessment of "good," "medium," "bad," or "scotomatous" retinal sensitivity.
   (c) *Scotoma finder*: Suprathreshold strategy that projects only 0 dB stimuli. Designed for patients with severely affected central vision and provides information concerning the scotoma.
   (d) *Fixation only*: For detecting the PRL.
3. Selecting "Grid Position" allows the position of the testing grid to be moved.
4. Custom test points can be added/removed using the "Add/Delete Stimuli" button.

3.9 Setting Up Fixation Target

1. After selecting start, the machine will perform automatic alignment, and the white light in the middle of the fixation circle will disappear.
2. Following the completion of automatic alignment, the operator can utilize manual controls to fine-tune the focus, brightness levels, and alignment (Fig. 2).
   (a) The default fixation target is the central target. This can be readjusted by the examiner.
3. Once the examiner is satisfied that the autofocused image is clear and that the patient can see the red circle fixation target, select "OK" and the camera will capture a retinal image for reference during the test.
4. Check the reference picture acquired. If the image is acceptable and in focus, select yes. Otherwise, select no, and the machine will prompt with an option to rerun autofocus or adjust using manual focusing.

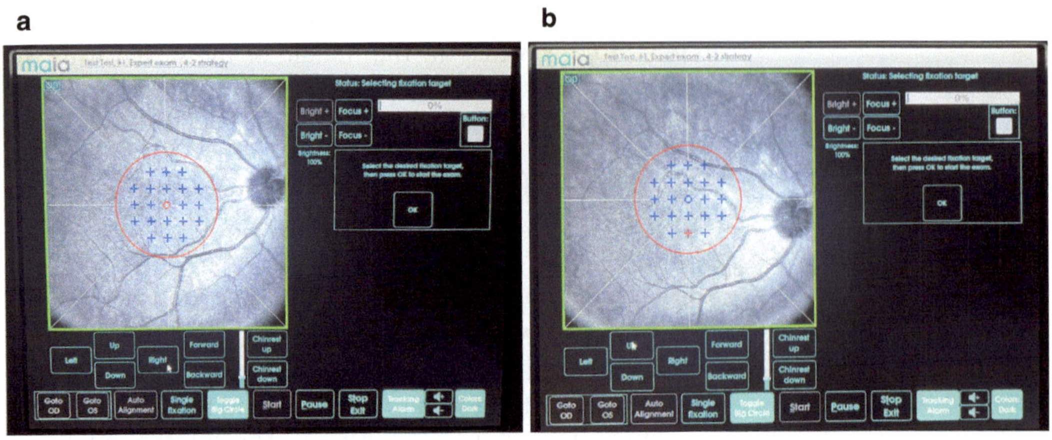

Fig. 2 After automatic alignment, the fixation target can be manually adjusted by the examiner. (**a**) Default central fixation target; (**b**) fixation target moved to test a retinal location superior to the fovea

3.10 Blind Spot Selection

1. The machine will then prompt the examiner to locate the blind spot by manually selecting the center of the optic disc on the screen. Incorrect identification of the disc's center can affect blind spot monitoring.
2. Once the optic disc has been selected, click "OK" to begin the test.

3.11 Commencing the Test

1. The patient is instructed to keep fixating on the center of the red circle and to press the button on the buzzer once they see a white flashing light.
2. When testing begins, the instrument will calibrate the patients' PRL (*see* **Note 8**). This usually takes around 10–15 s.
3. It may be helpful to encourage the patient to keep looking at the center as steadily as possible for the first 10–15 s, while the machine calibrates their PRL (*see* **Note 9**), and briefly count them down to when the first stimuli are presented.
4. Test stimuli will begin to be presented, with the first few being suprathreshold and the subsequent stimuli dimmer.

3.12 Monitoring During the Test

1. During the test, continuously remind the patient to keep fixating at the center of the red circle and to blind normally, as needed. The patient should also be reminded to not look for the flashing lights.
2. Depending on the individual and the condition, some features might require specific monitoring by the examiner. For example, patients with central scotomas, such as those with macular dystrophies, may alter their fixation more often so eye movement will require greater attention than other factors.
3. The testing instrument provides visual and auditory feedback on alignment and fixation monitoring. However, the examiner

should provide continuous monitoring throughout the test for fixation stability and blind spot mapping and to optimize alignment.

3.12.1 Test Progress

1. The camera image on the operator screen will provide information on the eye position and eye movement in real time.
2. The individual points being tested are shown in real time and when the response button is pressed and registered.
3. Test progression in terms of points with convergence of the staircase (i.e., determined threshold) is reported as a percentage, allowing positive encouragement about progress to be reported to the patient.
4. Just as patients may experience fatigue, some patients may require frequent updates on the progress of their test. Providing continuous encouragement and real-time updates during the testing process can keep the patient engaged and help re-establish concentration.

3.12.2 False Positives

1. False-positive responses to catch trials are evaluated by stimuli presented at the blind spot. These are presented approximately one per minute.
2. False-positive checks should be monitored for responses to stimuli presented at the optic disc (within the blue diamond; Fig. 3).
3. Strategies to minimize false positives include:
   (a) Providing ongoing instruction to ensure that the patient stays fixated on the central target and is not looking around at the testing spots.
   (b) Ensuring that there is no head tilt bringing the marked blind spot into view.
   (c) Ensuring there is sufficiently accurate fundus tracking.

3.12.3 Troubleshooting

1. If the eye position falls outside of tolerance, the test will pause, and an auditory cue will sound (Fig. 3).
2. The following components can be checked to troubleshoot the eye position:
   (a) A head tilt resulting in noncontact with the forehead rest.
   (b) Poor patient positioning or setup, such as patient slouching or straining, making it difficult to stay in the same position.
   (c) Poor tear film stability or incorrect fundus focus resulting in a change in accommodative effort of the eye.
   (d) Patient fatigue, requiring frequent breaks or re-establishing of concentration.

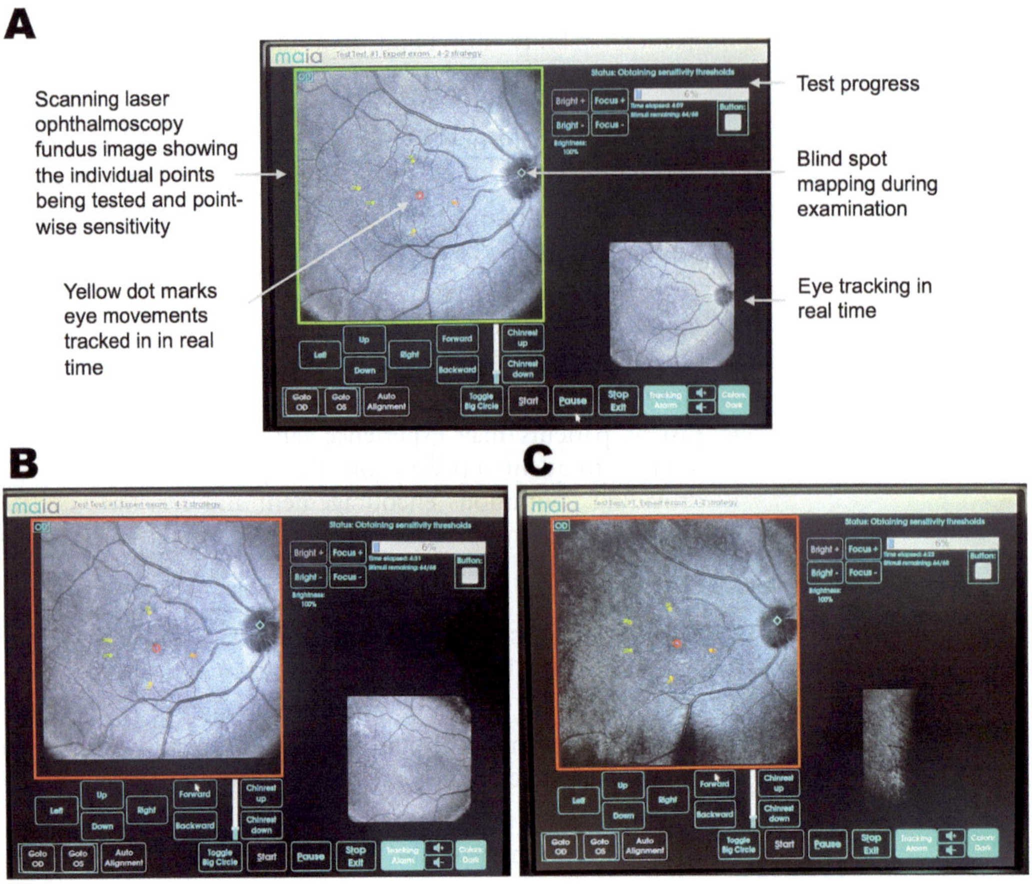

Fig. 3 (**a**) Fundus-tracked perimetry testing performed using the Macular Integrity Assessment (MAIA, CenterVue SpA., Icare Finland). (**b** and **c**) Red box around the scanning laser ophthalmoscopy image showing losses of fixation. Black edges or vignetting around the fundus image shown in real time indicates that the eye or the patient has moved

3.13 Finishing the Test and Exporting the Results

The results can be exported into various formats including:

1. A PDF printout of the eye report (Fig. 4):
   (a) This can be exported while on the results screen, using the "Print …" button and saving the report to a selected external USB.
2. PNG image of the IR and sensitivity data:
   (a) This can be exported while on the results screen, using the "Export" button and saving the image to a selected external USB.
3. Text file containing the threshold and fixation raw data. This can be exported from the System tab in Settings, using the "Raw Data Export" option.

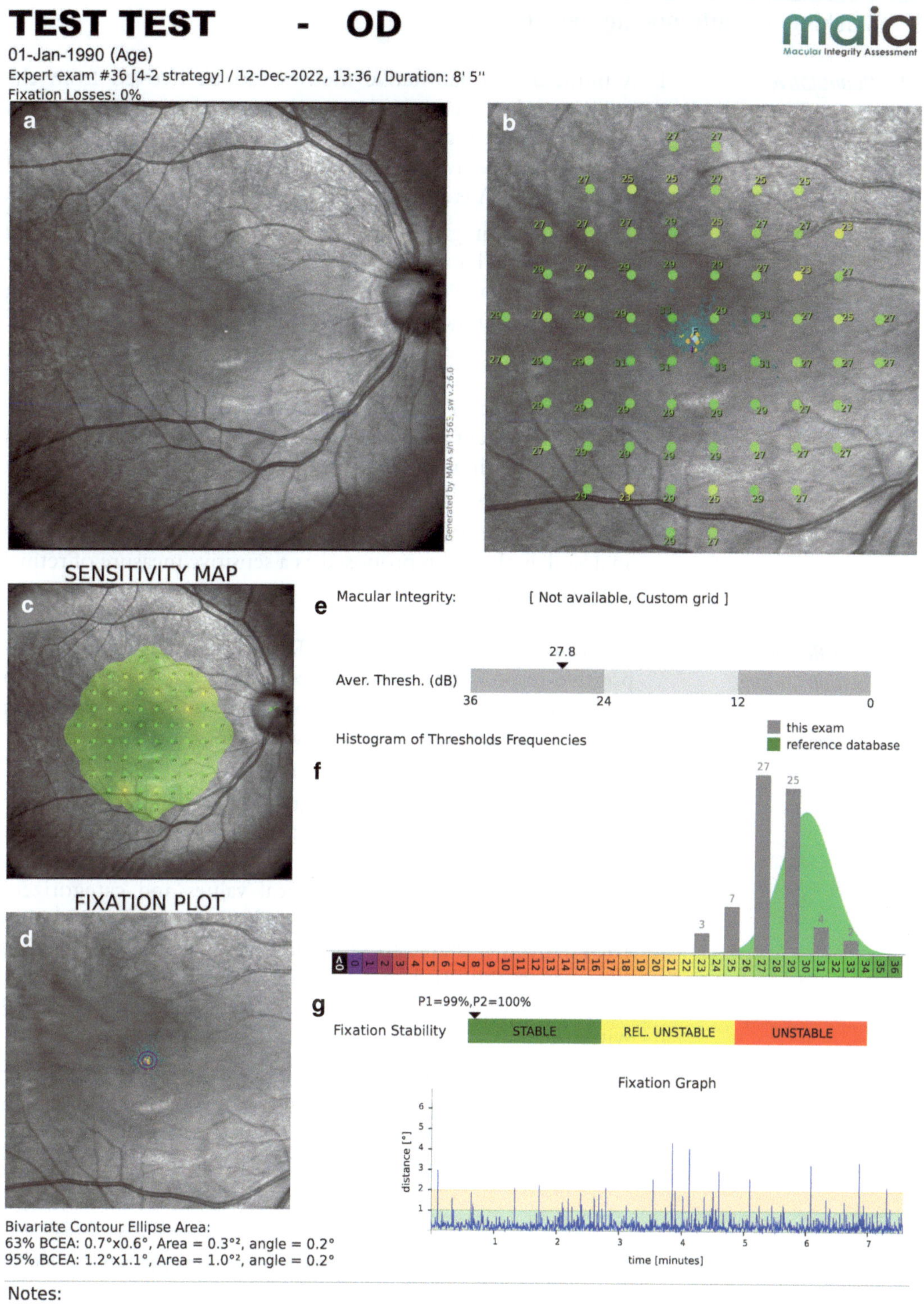

Fig. 4 Example of a PDF printout from the Macular Integrity Assessment (MAIA, CenterVue SpA., Icare Finland), taken from a healthy individual with no retinal disease. (**a**) Image of the fundus captured using scanning laser ophthalmoscopy (SLO). (**b**) Zoomed SLO fundus image showing individual sensitivity values at each testing point and the preferred retinal locus (PRL). (**c**) Interpolated "heat map" showing retinal sensitivity

4 Checking and Interpreting Results

4.1 Quantitative Metrics

1. A number of quantitative metrics can be derived from the fundus-tracked perimetry measurements; for example, mean sensitivity showing the mean of all points in a test grid or point-wise sensitivity. An example of the metrics exported from the MAIA is shown in Fig. 4.
2. The macular integrity index: A numerical value that describes the likelihood that a patient's responses are normal, suspect, or abnormal when compared to age-adjusted normative data; higher values suggest a greater likelihood of abnormal findings. It does not represent the severity of the disease process, and it does not have a direct relationship with the mean threshold sensitivity value. The macular integrity index is unique to the MAIA instrument and is only available when using the standard expert test setting (37 test points covering central 10°), using the full-threshold 4–2 strategy.
3. Three-dimensional visualization and volumetric sensitivity analysis has also been proposed as a sensitive measure of retinal function [23].

4.2 Test Reliability and Fixation Metrics

1. *Fixation losses:* Using the MAIA, fixation losses are false-positive responses to stimuli presented at the blind spot. An examination is considered reliable if the number of fixation losses exceeds an upper limit of acceptance, often taken to be between 25% and 33%, depending on the protocol [24].
2. *Fixation stability:* The fixation graph (Fig. 4g) describing the amplitude of eye movements over time and the percentage of fixation points located within a 1° (P1) and 2° (P2) radius distance are provided in numerical values and categorized. Stable fixation indicates >75% of fixation points are within P1; relatively unstable fixation indicates that <75% of fixation points are within P1, but >75% of fixation points are within P2; unstable fixation indicates <75% of fixation points are located within P2.
3. *Bivariate contour ellipse area (BCEA):* The BCEA is another method of assessing fixation stability. The ellipse of the BCEA encompasses the proportion of the fixation points (63% or 95%) used by the patient during the test, calculated in degrees^2.

Fig. 4 (continued) superimposed over the SLO fundus image. (**d**) Fixation plot over the zoomed SLO fundus image and PRL identification. (**e**) Macular integrity index and average sensitivity threshold. In standard grids, these may be color coded relative to the corresponding normal ranges in the MAIA reference database. (**f**) Histogram showing frequency of patient's threshold values (gray) compared with normal distribution (green) from the MAIA reference database. (**g**) Fixation graph describing the amplitude of eye movements over time

A smaller BCEA value implies a more stable fixation. The test is deemed unreliable if 95% BCEA is >4 degrees^2 in healthy individuals and >30 degrees^2 in individuals with retinal diseases [25].

5 Notes

1. Minimum pupil dilation: Several fundus-tracked perimetry devices have described a minimum pupil diameter in their technical specifications. For the MAIA device, minimum pupil diameter is specified as 2.5 mm and for the MP-3 device it is 4 mm [21]. In prior studies assessing retinal sensitivity under mesopic conditions, pupil dilation did not affect the consistency of the test results [26].
2. For fixed test patterns, point stimuli are presented in a testing grid with pre-specified testing points. A number of fixed test grids are available as standard on the instruments; for example, the rectilinear 10–2 grid (*see* Subheading 4). Custom tests can easily be created and uploaded into most fundus-controlled perimetry devices (*see* **Note 3**). Patient-tailored patterns have also been used to examine specific retinal areas; for example, a customized grid was used to cover the island of functional retina in a choroideremia gene therapy clinical trial [13].
3. For clinical trials, a customized grid is sometimes designed. Regulatory bodies such as the Food and Drug Administration (FDA) do not current dictate the type of grid that can be used for fundus-tracked perimetry but state that improvement of at least 7 dB needs to be seen in at least five pre-defined loci for significant clinical change.
4. Generally, fundus-tracked perimetry should be performed before exposing the retina to bright lights (e.g., retinal imaging). After exposure to lower levels of ambient light, additional adaptation time is not required, as it does not improve testing performance [27]. However, if fundus-tracked perimetry testing is performed after exposure to a bright light stimulus (e.g., slit lamp examination or ocular imaging), then 10 min of dark adaptation is recommended [27].
5. In contrast to the MAIA, the MP-3 allows for both mesopic (1.27 cd/m^2) and photopic (10 cd/m^2) testing. There are now also fundus-tracked perimetry instruments designed for testing in scotopic luminance (e.g., scotopic MAIA) that presents stimuli in two different wavelengths, which additionally enables the ability to distinguish between rod and cone sensitivities.

6. Ocular lubricants should be offered to the patient before the test or during breaks if they have clinical signs of a compromised tear film or symptoms of dry eye disease.
7. Not to be confused with the 10–2 grid. The "4–2" projection strategy describes the staircase psychophysical method, where light intensity is changed in 4 dB steps until there is a change from not seen to seen (or from seen to not seen) and then in 2 dB steps until the response is reversed. The "10–2" is the grid pattern covering 10° radius (i.e., the central 20° diameter) with 2° spacing between test points.
8. The preferred retinal loci (PRL) is the retinal area the patient uses to attempt fixation by attempting to "look at" a selected visual target. The PRLs provide information on the location and stability of a subject's fixation and provide a target for biofeedback training. On the MAIA, fixation is assessed by tracking eye movements at 25 times/s, and the result is plotted over the SLO fundus image (Fig. 4b and d), where the overall cluster of points shown describes the PRL. The MAIA identifies two main PRL reference points: the first PRL_initial is found after the initial 10 s of the examination and defines the center of the MAIA stimuli grid. The second PRL_final is found at the end of the examination and serves as the reference point to calculate fixation stability. Patients with stable fixation will present both PRLs in the same anatomical location.

Conflict of Interest

The authors have no conflicts of interest to declare that are relevant to the content of this chapter.

References

1. Saßmannshausen M, Zhou J, Pfau M, Thiele S, Steinberg J, Fleckenstein M, Holz FG, Schmitz-Valckenberg S (2021) Longitudinal analysis of retinal thickness and retinal function in eyes with large Drusen secondary to intermediate age-related macular degeneration. Ophthalmol Retina 5(3):241–250. https://doi.org/10.1016/j.oret.2020.07.019
2. Wu Z, Luu CD, Hodgson LAB, Caruso E, Brassington KH, Tindill N, Aung KZ, Harper CA, Wickremasinghe SS, Sandhu SS, McGuinness MB, Chen FK, Chakravarthy U, Arnold JJ, Heriot WJ, Durkin SR, Wintergerst MWM, Gorgi Zadeh S, Schultz T, Finger RP, Cohn AC, Baglin EK, Sharangan P, Guymer RH (2019) Secondary and exploratory outcomes of the subthreshold nanosecond laser intervention randomized trial in age-related macular degeneration: a LEAD study report. Ophthalmol Retina 3(12):1026–1034. https://doi.org/10.1016/j.oret.2019.07.008
3. Boned-Murillo A, Diaz-Barreda MD, Ferreras A, Bartolomé-Sesé I, Orduna-Hospital E, Montes-Rodríguez P, Ascaso J, Pinilla I (2021) Structural and functional findings in patients with moderate diabetic retinopathy. Graefes Arch Clin Exp Ophthalmol 259(12):3625–3635. https://doi.org/10.1007/s00417-021-05277-y
4. Vujosevic S, Midena E, Pilotto E, Radin PP, Chiesa L, Cavarzeran F (2006) Diabetic macular edema: correlation between microperimetry and optical coherence tomography findings.

Invest Ophthalmol Vis Sci 47(7):3044–3051. https://doi.org/10.1167/iovs.05-1141
5. Lima VC, Prata TS, Moraes CGVD, Kim J, Seiple W, Rosen RB, Liebmann JM, Ritch R (2010) A comparison between microperimetry and standard achromatic perimetry of the central visual field in eyes with glaucomatous paracentral visual-field defects. Br J Ophthalmol 94(1):64. https://doi.org/10.1136/bjo.2009.159772
6. Matsuura M, Murata H, Fujino Y, Hirasawa K, Yanagisawa M, Asaoka R (2018) Evaluating the usefulness of MP-3 microperimetry in glaucoma patients. Am J Ophthalmol 187:1–9. https://doi.org/10.1016/j.ajo.2017.12.002
7. Britten-Jones AC, O'Hare F, Edwards TL, Ayton LN (2022) Victorian evolution of inherited retinal diseases natural history registry (VENTURE study): rationale, methodology, and initial participant characteristics. Clin Experiment Ophthalmol 50(7):768–780. https://doi.org/10.1111/ceo.14110
8. Nguyen XT, Talib M, van Schooneveld MJ, Wijnholds J, van Genderen MM, Schalij-Delfos NE, Klaver CCW, Talsma HE, Fiocco M, Florijn RJ, Ten Brink JB, Cremers FPM, Meester-Smoor MA, Ingeborgh van den Born L, Hoyng CB, Thiadens A, Bergen AA, Boon CJF (2021) CRB1-associated retinal dystrophies: a prospective natural history study in anticipation of future clinical trials. Am J Ophthalmol. https://doi.org/10.1016/j.ajo.2021.07.021
9. Lad EM, Duncan JL, Liang W, Maguire MG, Ayala AR, Audo I, Birch DG, Carroll J, Cheetham JK, Durham TA, Fahim AT, Loo J, Deng Z, Mukherjee D, Heon E, Hufnagel RB, Guan B, Iannaccone A, Jaffe GJ, Kay CN, Michaelides M, Pennesi ME, Vincent A, Weng CY, Farsiu S (2022) Baseline microperimetry and OCT in the RUSH2A study: structure-function association and correlation with disease severity. Am J Ophthalmol 244:98–116. https://doi.org/10.1016/j.ajo.2022.08.013
10. Britten-Jones AC, Jin R, Gocuk SA, Cichello E, O'Hare F, Hickey DG, Edwards TL, Ayton LN (2021) The safety and efficacy of gene therapy treatment for monogenic retinal and optic nerve diseases: a systematic review. Genet Med 24(3):521–534. https://doi.org/10.1016/j.gim.2021.10.013
11. Fischer MD, Ochakovski GA, Beier B, Seitz IP, Vaheb Y, Kortuem C, Reichel FFL, Kuehlewein L, Kahle NA, Peters T, Girach A, Zrenner E, Ueffing M, MacLaren RE, Bartz-Schmidt KU, Wilhelm B (2019) Efficacy and safety of retinal gene therapy using adeno-associated virus vector for patients with choroideremia: a randomized clinical trial. JAMA Ophthalmol 29:29. https://doi.org/10.1001/jamaophthalmol.2019.3278
12. Lam BL, Davis JL, Gregori NZ, MacLaren RE, Girach A, Verriotto JD, Rodriguez B, Rosa PR, Zhang X, Feuer WJ (2019) Choroideremia gene therapy phase 2 clinical trial: 24-month results. Am J Ophthalmol 197:65–73. https://doi.org/10.1016/j.ajo.2018.09.012
13. MacLaren RE, Groppe M, Barnard AR, Cottriall CL, Tolmachova T, Seymour L, Clark KR, During MJ, Cremers FP, Black GC, Lotery AJ, Downes SM, Webster AR, Seabra MC (2014) Retinal gene therapy in patients with choroideremia: initial findings from a phase 1/2 clinical trial. Lancet 383(9923):1129–1137. https://doi.org/10.1016/S0140-6736(13)62117-0
14. Cehajic-Kapetanovic J, Xue K, Martinez-Fernandez de la Camara C, Nanda A, Davies A, Wood LJ, Salvetti AP, Fischer MD, Aylward JW, Barnard AR, Jolly JK, Luo E, Lujan BJ, Ong T, Girach A, Black GCM, Gregori NZ, Davis JL, Rosa PR, Lotery AJ, Lam BL, Stanga PE, MacLaren RE (2020) Initial results from a first-in-human gene therapy trial on X-linked retinitis pigmentosa caused by mutations in RPGR. Nat Med 26(3):354–359. https://doi.org/10.1038/s41591-020-0763-1
15. Fischer MD, Michalakis S, Wilhelm B, Zobor D, Muehlfriedel R, Kohl S, Weisschuh N, Ochakovski GA, Klein R, Schoen C, Sothilingam V, Garcia-Garrido M, Kuehlewein L, Kahle N, Werner A, Dauletbekov D, Paquet-Durand F, Tsang S, Martus P, Peters T, Seeliger M, Bartz-Schmidt KU, Ueffing M, Zrenner E, Biel M, Wissinger B (2020) Safety and vision outcomes of subretinal gene therapy targeting cone photoreceptors in achromatopsia: a nonrandomized controlled trial. JAMA Ophthalmol 138(6):643–651. https://doi.org/10.1001/jamaophthalmol.2020.1032
16. Dimopoulos IS, Hoang SC, Radziwon A, Binczyk NM, Seabra MC, MacLaren RE, Somani R, Tennant MTS, MacDonald IM (2018) Two-year results after AAV2-mediated gene therapy for choroideremia: the Alberta experience. Am J Ophthalmol 193:130–142. https://doi.org/10.1016/j.ajo.2018.06.011
17. Bainbridge JWB, Mehat MS, Sundaram V, Robbie SJ, Barker SE, Ripamonti C, Georgiadis A, Mowat FM, Beattie SG, Gardner PJ, Feathers KL, Luong VA, Yzer S, Balaggan K, Viswanathan A, De Ravel TJL, Casteels I, Holder GE, Tyler N, Fitzke FW, Weleber RG, Nardini M, Moore AT,

Thompson DA, Petersen-Jones SM, Michaelides M, Van Den Born LI, Stockman A, Smith AJ, Rubin G, Ali RR (2015) Long-term effect of gene therapy on Leber's congenital amaurosis. New Eng J Med 372(20):1887–1897. https://doi.org/10.1056/NEJMoa1414221

18. Le Meur G, Lebranchu P, Billaud F, Adjali O, Schmitt S, Bezieau S, Pereon Y, Valabregue R, Ivan C, Darmon C, Moullier P, Rolling F, Weber M (2018) Safety and long-term efficacy of AAV4 gene therapy in patients with RPE65 leber congenital amaurosis. Mol Ther 26(1):256–268. https://doi.org/10.1016/j.ymthe.2017.09.014

19. Cukras C, Wiley HE, Jeffrey BG, Sen HN, Turriff A, Zeng Y, Vijayasarathy C, Marangoni D, Ziccardi L, Kjellstrom S, Park TK, Hiriyanna S, Wright JF, Colosi P, Wu Z, Bush RA, Wei LL, Sieving PA (2018) Retinal AAV8-RS1 gene therapy for X-linked retinoschisis: initial findings from a phase I/IIa trial by Intravitreal delivery. Mol Ther 26(9):2282–2294. https://doi.org/10.1016/j.ymthe.2018.05.025

20. Balasubramanian S, Uji A, Lei J, Velaga S, Nittala M, Sadda S (2018) Interdevice comparison of retinal sensitivity assessments in a healthy population: the CenterVue MAIA and the Nidek MP-3 microperimeters. Br J Ophthalmol 102(1):109–113. https://doi.org/10.1136/bjophthalmol-2017-310258

21. Pfau M, Jolly JK, Wu Z, Denniss J, Lad EM, Guymer RH, Fleckenstein M, Holz FG, Schmitz-Valckenberg S (2020) Fundus-controlled perimetry (microperimetry): application as outcome measure in clinical trials. Prog Retin Eye Res:100907. https://doi.org/10.1016/j.preteyeres.2020.100907

22. Fischer MD, Ochakovski GA, Beier B, Seitz IP, Vaheb Y, Kortuem C, Reichel FFL, Kuehlewein L, Kahle NA, Peters T, Girach A, Zrenner E, Ueffing M, MacLaren RE, Bartz-Schmidt K, Wilhelm B (2020) Changes in retinal sensitivity after gene therapy in choroideremia. Retina 40(1):160–168. https://doi.org/10.1097/IAE.0000000000002360

23. Josan AS, Buckley TMW, Wood LJ, Jolly JK, Cehajic-Kapetanovic J, MacLaren RE (2021) Microperimetry hill of vision and volumetric measures of retinal sensitivity. Transl Vis Sci Technol 10(7):12. https://doi.org/10.1167/tvst.10.7.12

24. Wu Z, Ayton LN, Guymer RH, Luu CD (2013) Intrasession test-retest variability of microperimetry in age-related macular degeneration. Invest Ophthalmol Vis Sci 54(12):7378–7385. https://doi.org/10.1167/iovs.13-12617

25. Josan AS, Farrance I, Taylor LJ, Adeyoju D, Buckley TMW, Jolly JK, MacLaren RE (2023) Microperimetry reliability assessed from fixation performance. Transl Vis Sci Technol 12(5):21. https://doi.org/10.1167/tvst.12.5.21

26. Han RC, Jolly JK, Xue K, MacLaren RE (2017) Effects of pupil dilation on MAIA microperimetry. Clin Experiment Ophthalmol 45(5):489–495. https://doi.org/10.1111/ceo.12907

27. Han RC, Gray JM, Han J, Maclaren RE, Jolly JK (2019) Optimisation of dark adaptation time required for mesopic microperimetry. Br J Ophthalmol 103(8):1092–1098. https://doi.org/10.1136/bjophthalmol-2018-312253

Chapter 16

Assessing Visual Function in Retinal Gene Therapy

Andrew J. Anderson, Michael Bach, and Ruth E. Hogg

Abstract

Conventional tests of visual function include measures such as visual acuity, contrast sensitivity, perimetry, and color vision. However, standard clinical tests are typically designed to test those with normal or near-normal sensitivities. Such tests may be inappropriate where profound vision loss exists, as may be the case in retinal gene therapy if the therapy is nascent and/or the therapy is unable to fully restore the visual system. Here we provide step-by-step instructions for performing a battery of three computerized tests to quantify light detection, light localization, and motion detection, in eyes with profound vision loss. The tests use conventional computer hardware, do not require extensive calibration, and are relatively short to perform.

Key words Vision testing, Visual function, Sensitivity, Detection, Localization, Motion, Retinal gene therapy

1 Introduction

Normal visual function consists of visually sensed variations in light across several domains. For example, the amount of light can be sensed (light detection) as can variations in the amount of light as a function of position (e.g., gross light localization, or more spatially fine grating resolution). Variations over time can also be sensed (temporal vision and motion detection), as can variations in the spectral composition of a light (color vision). More complex visual functions also exist, such as the grouping of several local variations in lights—potentially representing contours or features—into an object or using intraocular differences to generate a sense of stereoscopic depth. There is an almost inexhaustible array of potential dimensions across which the limits of visual function may be measured. For simplicity, here we concentrate on a subset of basic visual functions, being light detection and localization, and motion discrimination.

Many tests of visual function are designed to assess those with normal, or near-normal, sensitivities. However, it is possible that participants in trials of gene therapy will have profound levels of

Livia S. Carvalho (ed.), *Retinal Gene Therapy: Methods and Protocols*, Methods in Molecular Biology, vol. 3023,
https://doi.org/10.1007/978-1-0716-5198-8_16,
© The Author(s), under exclusive license to Springer Science+Business Media, LLC, part of Springer Nature 2026

vision loss, because the therapy is nascent and/or the therapy is unable to fully restore the visual system. It is instructive to consider pitfalls that have arisen recently when testing visual function in those with profound vision loss: people with visual prostheses (or "bionic eyes"). Criticism has been raised regarding the validity of some visual testing [1], which potentially allows participants to perform the tests without possessing the nominal visual function: for example, using time-of-arrival cues at a single location to predict motion direction [2] or head-scanning with a single light sensor to perform a resolution task [1]. In addition to issues of test construction, a review of 34 papers assessing visual outcomes for visual prostheses in retinitis pigmentosa found more general methodological errors, including a lack of masking of treatment allocation and a failure to fully assess visual status pre-intervention [2]. Such concerns have lead authors to question to what degree the visual restoration of spatial and temporal vision has been demonstrated by previous studies, rather than simply an improvement in light perception [1, 2]. It is with these recent cautionary tales in mind that we outline the following protocol for assessing visual function in gene therapy trials.

It is also worthwhile considering the distinction between the restoration of visual sensation and being able to perform a task typically done visually in a visually healthy observer. For example, navigating a crowded street is normally done visually but could also be done by a completely blind person using a device that provides audible navigation cues: such a device allows a task to be performed, although clearly it does not restore the visual sense. Well-designed tests will try to minimize the possibility that a person is using an extraneous cue—whether visual or other—to perform the task, although ruling out all such cues within a simple battery of widely accessible tests is likely to be challenging. The reader is encouraged to peruse recent commentaries on some of these challenges [1, 2].

Where the level of visual function permits, testing can be done using conventional tests and protocols, such a visual acuity [3], contrast sensitivity [4], and perimetry [5]. Protocols have been described elsewhere (including in this book, for assessing spatial extent of increment contrast sensitivity via fundus-tracked perimetry) and so are not considered further here. Rather, we consider those persons for whom conventional measures of visual acuity fail [6], being approximately where the upper line of a visual acuity chart cannot be read even with a reduced test distance of 1 m (=1/60 or logMAR 1.8). Such acuities fall within the "light perception" and "no light perception" categories of International Classification of Diseases (11th Revision). In this chapter, our protocol is based upon a previously described test battery (The "Basic Light, Location and Motion Test," BaLM) developed by Michael Bach and colleagues [7] which has been shown to quantitatively separate very low vision categories (hand movements, light perception, no light perception) across categories of light detection, along with the discrimination of direction, timing, and motion. We provide some modifications from the default test parameters originally suggested by these authors, primarily by shortening the time in which people have to respond and so minimizing (although not eliminating) strategies based on eye scanning. We also omit the previously described test component assessing temporal resolution (i.e., discriminating two light flashes separated by 200 ms) [7] which had been of greatest utility for low-temporal-resolution retinal implants. BaLM has previously been used as a secondary endpoint for assessing the performance of a retinal implant for restoring vision in end-stage retinitis pigmentosa [8].

When selecting a protocol, it is important to consider the appropriateness of the tests for the participant group. This may include considering any test endpoints to be used to establish a therapy's efficacy and the acceptability of these to agencies regulating new therapies (e.g., the Food and Drug Administration in the USA). Visual acuity was the most common clinical trial endpoint in a recent review of inherited retinal disease therapy trials, with perimetry and tests of mobility also commonly used [9]. However, limitations in using acuity as a measure were also noted, along with the need to have measures that have been tailored to the level of vision loss [9]. Our protocol is designed for those with profound vision loss and so for whom visual acuity testing and conventional tests of peripheral vision (e.g., perimetry) are inappropriate. To make the protocol accessible to a wide range of clinicians and researchers, it uses free software which can be run on conventional computer hardware, with the additional of some simple equipment (e.g., neutral density filters).

2 Materials

1. The online $FrACT_{10}$ testing suite (https://michaelbach.de/fract/), run in fullscreen ("Fullscr.") mode. This suite contains a module $BALM_{10}$, whose tests are based on the original BaLM software (https://michaelbach.de/sci/stim/balm/) [7] but require no software to be installed. Note that stimuli generated by $BaLM_{10}$ are reasonably straightforward (Fig. 1) and so could also be replicated within other visual presentation software, such as the open-source program PsychoPy [10].
2. A computer monitor for displaying stimuli, whose viewable screen dimensions are at least 31 cm horizontally and vertically (*see* **Note 1**). The monitor should be able to produce a sustained, full-screen luminance of at least 320 cd/m^2.

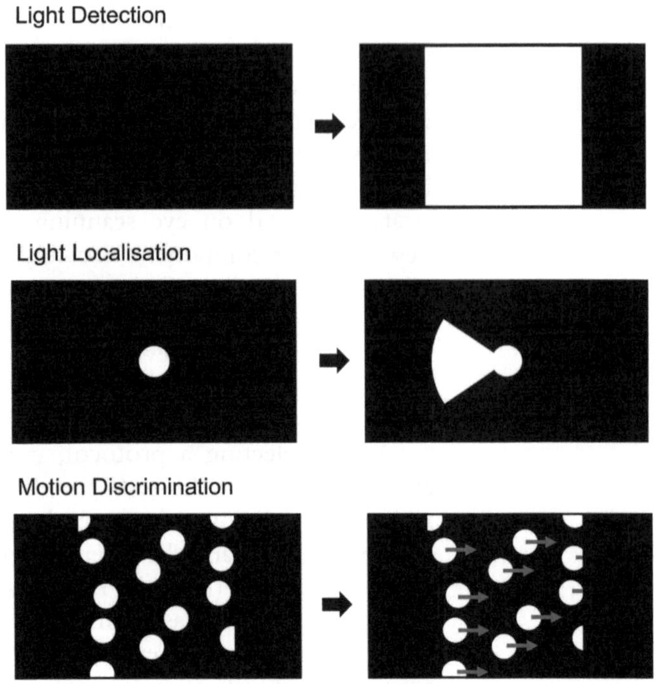

Fig. 1 Illustration of the three tests, representing the screen before (left panels) and during (right panels) a stimulus presentation. For the light detection thresholds (upper panels), participants indicate whether a large flash (30° diameter square, 0.2 s duration) was presented. The flash luminance is adjusted in 0.5 log unit steps to determine threshold, by viewing the screen through neutral density (ND) filters. For light localization (middle panels), participants indicate the direction (up, down, left, right) of a wedge of light (of angle 66.6° and extending 15° from fixation), presented for up to 2 s. For motion discrimination, participants indicate the direction of drift (up, down, left, right) of a high contrast field of dots (3.3° diameter, drifting at 3.3°/s in a 30° diameter square field), presented for up to 2 s; note that the gray arrows are to illustrate the drift direction and do not appear on the screen. For both the light localization and motion discrimination tests, participants view the screen through an ND filter 0.5 ND lower than the light detection threshold

3. A photometer for measuring screen luminance. $BaLM_{10}$ stimuli are high contrast (white on black), and so linearization of the monitor's output ("gamma correction" [11]) is not required. As such, a low-cost photometer (e.g., one of the Spyder range of screen calibrators, from Datacolor (https://www.datacolor.com)) is sufficient.
4. A chin and forehead rest. For simplicity, hereafter we simply refer to this combination as a chin rest.
5. Alcohol for disinfecting the chin rest.
6. An opaque eye patch that can form a light-tight seal, to test each eye individually.

7. Neutral density (ND) filters, in 0.5 steps from 0.5 to 4.5 ND. Filters need to be appropriately mounted so that they can be placed securely in front of one eye and the entirety of the screen viewed but that stray light from around the filter is avoided. Close-fitting welding goggles can be used to achieve an appropriate seal, and gelatin ND filters can typically be cut to fit into the aperture of welding goggles and are relatively cheap. ND values give how much, in $\log_{10}$ units, a filter reduces the amount of light passing through it. For example, 0 ND transmits all light ($10^0 = 1$) and so is equivalent to no filter being in place, with an ND of 2.0 reducing the light 100-fold ($10^2 = 100$) and an ND of 3.0 reducing the light 1000-fold ($10^3 = 1000$). This logarithmic progression also means that ND filters are additive: For example, a 3.5 ND filter (reducing light 3162-fold) can be made by stacking a 1.0 ND and a 2.5 ND filter (reducing light tenfold at the first filter and a further 316.2-fold at the second).

3 Methods

3.1 Test Preparation

1. Prior to first testing, the monitor should have its brightness and contrast settings adjusted so the luminance is 320 cd/m^2 for a full-screen white image (RGB = 1.0, 1.0, 1.0). Ideally, the luminance of a full-screen black image (RGB = 0.0, 0.0, 0.0) should be 3.2 cd/m^2 or less, to ensure a high contrast ($\geq$99%) for the black and white test images. Although 1.2 log units lower than the projector-produced luminance of 5100 cd/m^2 used by Bach et al. [7], half of participants in their study classified as having "light-perception" vision had light detection thresholds closer to 100 cd/m^2 or better. Testing 0.5 log units above this threshold requires a luminance of 320 cd/m^2. A luminance of 320 cd/m^2 is also within the range of luminances recommended for visual acuity testing [12].
2. The "Settings" option of FrACT$_{10}$ should be selected, and, for size calibration, the length of the blue calibration bar measured and entered within the "General" tab. Alternatively, FrACT$_{10}$ provides an option for size calibration using a credit card or similar plastic card for reference ("Calibrate with plastic card (e.g., credit card)" button).
3. Select the "BaLM" tab. While the default parameter values should match those listed below, it is advisable to check that this has occurred, particularly for users unfamiliar with the software. Default BaLM parameter values can be restored by selecting "BALM$_{10}$" in the "PRESETS" menu.
4. Set the following "General" parameters:

- "ISI [milliseconds]" = 500
- "Timeout [seconds] = Max. reaction time" = 2
- "Distance [centimeters]" = 57
- "Total extent [°]" = 15

5. Set the following "Light" parameters:
   - "Light On [ms]" = 200
   - "# trials" = 24
6. Set the following "Location" parameters:
   - "Diameter [°]" = 5
   - "# trials (same as Motion)" = 20
   - "Eccentricity [°]" = 15
7. Set the following "Motion" parameters:
   - "Diameter [°]" = 3.3
   - "# trials (same as Location)" = 20
   - "Speed [°/s]" = 3.3
8. Click "OK" to return to the $FrACT_{10}$ home screen.
9. The chin rest should be placed 57 cm from the monitor. Each centimeter on the screen subtends 1 degree at this distance, making it easy to directly confirm with a ruler that the angular size of stimuli is as expected (e.g., a target designed to subtend 5 degrees will be 5 cm on the screen). Refractive correction for this working distance is also easy (= 1/0.57 = 1.75 dioptres).
10. Check that the computer's volume is appropriately set so that an observer could hear both the auditory tones indicating a trial in BaLM and any auditory feedback.

3.2 Test Protocols

1. Typically, testing should be done monocularly, with the non-tested eye patched or—if the goggles allow—occluded. Where there is likely to be a large asymmetry in sensitivity between eyes (e.g., in uniocular disease or in a therapy applied to one eye) (*see* **Note 2**), obtaining a light-tight seal about the non-tested eye is particularly important to prevent stray light to the fellow eye acting as a cue. Taping the eyelids closed behind the patch might be appropriate in such circumstances.
2. Testing should be performed in a quiet room, free of distractions.
3. Room lights should be on initially, to help the participant safely navigate the room. They can remain on during periods of instruction or when physical manipulations need to be being performed (e.g., adjusting seats and chin rests and changing filters in the goggles). Room lights should be turned off during testing. Room lights should be turned on again after testing, to help the participant safely exit the room.

4. Tests should be performed in the order given below.
5. Any required refractive correction should be placed in a trial frame, with appropriate correction for the working distance (57 cm = 1.75 D), as appropriate. Although large, $BaLM_{10}$ stimuli are hard-edged, and contrast sensitivity to edges decreases by around half (0.3 log units) with each dioptre of defocus, albeit in normally sighted observers [13]. Given that subjective refraction will likely be impossible, objective methods (e.g., retinoscopy, autorefraction, ocular biometry, previous records of refractive error) may be required to estimate the refractive error, at least to within a few dioptres. At the very least, high refractive error should be ruled out as a potential cause of decreased visual sensitivity, particularly given that refractive error is associated with some eye diseases, such as inherited retinal disease [14].

3.2.1 Light Detection Threshold

1. This test measures the participant's sensitivity for detecting a large, uniform flash of light. Sensitivity is measured by determining the lowest luminance flash that can be reliably detected. The test has similarities with the full-field stimulus threshold (FST) test [15] but with coarser resolution for final thresholds and with better control for participants guessing.
2. Initially place no filter (0ND) in the goggles for the test eye, and patch the untested eye (*see* **Note 3**).
3. With assistance, if necessarily, have the participant place their index and ring fingers of their right hand on the left and right arrow keys of the computer keyboard, respectively. Check the participant's familiarity with the arrow keys and that they are able to locate and push each key (left arrow, index finger; right arrow, ring finger; up and down arrows, middle finger) without the need to reposition their hand or look at the keyboard.
4. *Instructions*: "Please look straight ahead at the monitor in front of you. Each test trial is indicated by an auditory tone. If you see a large flash on the screen that accompanies the tone, press the RIGHT arrow, and if you don't see a flash accompanying the tone, press the LEFT arrow. You have up to 2 s to respond before the next trial begins. Make sure you continue to blink during the test (*see* **Note 4**). A good time to blink is each time you push a key—in this way, you won't miss any flashes. The computer will also provide feedback, with a high tone for correct responses and a low tone for incorrect responses" (*see* **Note 5**).
5. Commence the test by selecting the $BaLM_{10}$ button, followed by the "Light ('1')" button.

6. For participants with hearing loss that may make listening for the trial tone unreliable (e.g., in the genetic disorder Usher syndrome), the instructor can tap the participant on the upper arm each time the tone sounds to indicate each trial.
7. The participant should achieve a mean hit rate of at least 75% (18/24 correct, or better) on their initial training run. If this is not achieved, *reinstruct* the observer, and rerun the test. If the participant still fails to reach 75% correct, then the result is recorded as "no light perception" (NLP). No further testing is performed, either for this test or for the test components described subsequently (*see* **Note 6**).
8. Assuming the participant's response has not been recorded as "NLP," repeat the test with the luminance decreased in 1.5 ND steps until 75% correct is no longer achieved, and then increase the luminance in 0.5 ND steps (to a maximum of two) until 75% correct is achieved. The highest ND where a mean hit rate of at least 75% correct is reached is the participant's light detection threshold. To avoid sudden alterations in the adaptation state of the eye, participants should be instructed to close their eyes while the goggles are removed and filters altered.

3.2.2 Light Localization

1. This test determines the participant's ability to localize where a wedge-shaped patch of light appeared. Performance is measured as the proportion of times the participant can localize the patch.
2. Patch the non-tested eye.
3. Test with a luminance 0.5 ND lower (i.e., a luminance 3.12× higher) than the above light threshold for the test eye (e.g., if the light threshold was 3.5ND, location testing should be done with the participant viewing through a 3.0ND filter). If the light threshold is 0ND, testing should be done at 0ND (i.e., no filter).
4. *Instructions*: "Please look straight ahead at the central circle. Each test trial is indicated by an auditory tone, in which a large wedge of light appears around the central circle. Press the UP, DOWN, LEFT, or RIGHT arrow key that corresponds to the location of the wedge of light. Wedges only appear for a maximum of 2 s, and you must respond prior to the segment disappearing. Keep looking at the central circle at all times during the test, and remember to blink regularly. A good time to blink is each time you press an arrow key."
5. Commence the test by selecting the $BaLM_{10}$ button, followed by the "Location ('2')" button.
6. Perform one test for familiarization and then one for the final test result (*see* **Note 7**). Participants are considered to have been able to localize the light at above chance levels if their

mean hit rate is greater than 65% (13/20 correct), a discrete approximation of the 62.5% used previously by Bach et al. Hit rates progressively higher than this threshold represent progressively increased test performance.

7. Where vision loss, or the visual restoration, is not expected to be spatially uniform (e.g., therapy is applied to a localized quadrant of the retina), analyzing the hit rate for each light wedge direction separately may be appropriate. To access this information, click the "Settings" button and, on the "General" tab, select the "Full history" option for "Results to clipboard."
8. When testing the second eye, the familiarization test may be skipped.

3.2.3 Motion Discrimination

1. The test measures the participant's ability to determine the direction of motion of a large, slowly drifting field of dots. Performance is measured as the proportion of times the participant can correctly give the motion direction.
2. Patch the non-tested eye.
3. Conduct the test using the same ND filter as that used to test the eye in "Light Localization."
4. *Instructions*: "Each time the auditory tone sounds, you will be shown a field of white dots on the screen. The field will move slowly either upwards, downwards, to the left, or to the right. It only moves for 2 s, and you must indicate the direction of the movement with a corresponding arrow-key press (UP, DOWN, LEFT, or RIGHT) before the motion stops. Remember to blink regularly. A good time to blink is each time you press an arrow key."
5. Commence the test by selecting the $BaLM_{10}$ button, followed by the "Motion ('3')" button.
6. Perform one test for familiarization, followed by one for the final test result. Participants are considered to have been able to discriminate motion at above chance levels if their mean hit rate is greater than 65% (13/20 correct), a discrete approximation of the 62.5% used previously by Bach et al. Hit rates progressively higher than this threshold represent progressively increased test performance.
7. When testing the second eye, the familiarization test may be skipped.

3.3 Documentation

1. $BaLM_{10}$ does not provide any long-term storage of test results but does allow all export options available in $FrACT_{10}$. This exported data can be used to investigate hit rates as a function of visual field quadrant in the light localization test if the "Full history" (under Settings>General>Results to clipboard) is selected, as noted previously. Multiplying the number giving

the quadrant direction by 45 will give the quadrant angle in degrees, proceeding anticlockwise from the right (=0).

2. Testers are encouraged to manually record hit rates on a participant running sheet designed for the particular study or clinical needs. Such contemporaneous recording will ensure results are available in the event of computer or printing errors, as well as providing a method for tracking how much of the protocol has been completed. Such tracking is particularly useful should testing be interrupted by scheduled breaks or by unanticipated disruptions.
3. Testers should perform a complete run of any test protocol they develop and check that all required results can be exported in the format they require, prior to testing actual participants.

4 Notes

1. Each test in our $BaLM_{10}$ protocol is nominally confined to the central 30° (diameter) of the visual field. However, this will only be true if participants can fixate upon the center of the screen, which is likely to be difficult for persons with profoundly reduced vision. In addition, our use of goggles with ND filters means that a participant's fixation cannot be directly monitored by the tester, as might be done when testing without goggles [16]. Therefore, testers should be aware that, although $BaLM_{10}$ tests a 30° field, the location of this field on the participant's retina cannot typically be known accurately.
2. Where possible, the person administering the test should be blinded to the observer's treatment status (therapy/no gene therapy).
3. When testing each eye separately, what order each eye is tested in should be considered. Always testing with a particular eye first (e.g., the right) is simple, although it does mean that the occluding patch must be moved after each test run and that ND filters—if different between the two eyes—must be frequently changed. Testing one eye first on all tests, followed by the other, eliminates these inconveniences, although it does increase the likelihood of ordering effects—either through fatigue or learning—between the eyes. That the tests are short, and that a familiarization run is given between tests, should help mitigate fatigue and learning effects. Providing appropriate breaks for participants can also reduce fatigue effects. In the context of a clinical trial, ordering effects might also be appropriately controlled by randomizing or counterbalancing the eye tested between participants and/or participant visits. Ultimately, there is no single correct approach, and

testers are encouraged to consider in advance what test ordering will best serve the needs of their study and/or the needs of their individual participant.

4. Participants can get dry eyes during observation tasks, likely due to a reduced blink rate and the eyelids being more widely open in upgaze while staring at screens [17].
5. In the tests described above, participants are forced to make a response within a relatively short time. This has the potential to produce anxiety in participants, particularly if they are concerned that they will get the test "wrong" and that this might affect aspects of their treatment or care. Testers should therefore try to foster a friendly and relaxed atmosphere. They should reassure participants that a certain number of "wrong" responses are expected in each test and that these help the computer determine what can and cannot be seen. They should also be reassured that, even though sometimes it might feel that they are reduced to purely guessing, the computer can determine whether they are truly guessing or whether they are getting slightly more correct than might be expected based on guessing alone. The use of familiarization runs for each of the tests should help most participants feel comfortable with what is being asked of them. If scripts are desired, the following text can append the instructions given above:

   "Do not be anxious if sometimes your responses are incorrect, as this is expected and is part of the test determining what you can and cannot see. Even though sometimes it might feel that you are simply guessing, the computer can determine whether you are getting slightly more correct than might be expected based on guessing alone."
6. Extensive training on a test should be avoided as it will likely increase the risk of observers learning to do the task using extraneous cues [1].
7. If retesting a participant within a short period (e.g., a month or less since the last test), familiarization tests could be terminated after a few trials, assuming participant indicates they are happy with the nature of the test and the instructor is similarly happy.

Conflict of Interest

The authors have no conflicts of interest to declare that are relevant to the content of this chapter.

References

1. Peli E (2020) Testing vision is not testing for vision. Transl Vis Sci Technol 9(13):32
2. Hallum LE, Dakin SC (2021) Retinal implantation of electronic vision prostheses to treat retinitis pigmentosa: a systematic review. Transl Vis Sci Technol 10(10):8
3. Bailey IL, Lovie JE (1976) New design principles for visual acuity letter charts. Am J Optom Physiol Optic 53:740–745
4. Pelli DG, Robson JG, Wilkins AJ (1987) Designing a new letter chart for measuring contrast sensitivity. Clin Vis Sci 2(3):187–199
5. Heijl A, Patella VM, Bengtsson B (2012) The field analyzer primer: effective perimetry, 4th edn. Carl Zeiss Meditec, Inc, p 158
6. Schulze-Bonsel K, Feltgen N, Burau H, Hansen L, Bach M (2006) Visual acuities "hand motion" and "counting fingers" can be quantified with the freiburg visual acuity test. Investig Ophthalmol Vis Sci 47(3):1236–1240
7. Bach M, Wilke M, Wilhelm B, Zrenner E, Wilke R (2010) Basic quantitative assessment of visual performance in patients with very low vision. Investig Ophthalmol Vis Sci 51(2): 1255–1260
8. Edwards TL, Cottriall CL, Xue K, Simunovic MP, Ramsden JD, Zrenner E, MacLaren RE (2018) Assessment of the electronic retinal implant alpha AMS in restoring vision to blind patients with end-stage retinitis pigmentosa. Ophthalmology 125(3):432–443
9. Igoe JM, Lam BL, Gregori NZ (2024) Update on clinical trial endpoints in gene therapy trials for inherited retinal diseases. J Clin Med 13(18)
10. Peirce JW et al (2019) PsychoPy2: experiments in behavior made easy. Behav Res Methods 51(1):195–203
11. Metha AB, Vingrys AJ, Badcock DR (1993) Calibration of a color monitor for visual psychophysics. Behav Res Methods Instrum Comput 25(3):371–383
12. Sheedy JE, Bailey IL, Raasch TW (1984) Visual acuity and chart luminance. Am J Optom Physiol Optic 61(9):595–600
13. Jansonius NM, Kooijman AC (1997) The effect of defocus on edge contrast sensitivity. Ophthalmic Physiol Opt 17(2):128–132
14. Yassin SH et al (2025) Refractive error in inherited retinal disease. Am J Ophthalmol 269: 381–392
15. Shi LF, Hall AJ, Thompson DA (2024) Full-field stimulus threshold testing: a scoping review of current practice. Eye (Lond) 38(1): 33–53
16. Sepulveda JA, Wood JM, Lacherez P, Anderson AJ, McKendrick AM (2023) The relationship between central and mid-peripheral motion perception and the hazard perception test in younger and older adults. Ophthalmic Physiol Opt 43(5):1211–1222
17. Blehm C, Vishnu S, Khattak A, Mitra S, Yee RW (2005) Computer vision syndrome: a review. Surv Ophthalmol 50(3):253–262

Index

Livia S. Carvalho (ed.), *Retinal Gene Therapy: Methods and Protocols*, Methods in Molecular Biology, vol. 3023,
https://doi.org/10.1007/978-1-0716-5198-8,
© The Editor(s) (if applicable) and The Author(s), under exclusive license to Springer Science+Business Media, LLC, part of Springer Nature 2026

MIX
Papier aus verantwortungsvollen Quellen
Paper from responsible sources
FSC® C105338

If you have any concerns about our products,
you can contact us on
ProductSafety@springernature.com

In case Publisher is established outside the EU,
the EU authorized representative is:
Springer Nature Customer Service Center GmbH
Europaplatz 3, 69115 Heidelberg, Germany

Printed by Libri Plureos GmbH
in Hamburg, Germany